QUANTUM CHEMISTRY WORKBOOK

QUANTUM CHEMISTRY WORKBOOK

Basic Concepts and Procedures in the Theory of the Electronic Structure of Matter

Jean-Louis Calais
University of Uppsala
Uppsala, Sweden

A Wiley-Interscience Publication
JOHN WILEY & SONS, INC.
New York   Chichester   Brisbane   Toronto   Singapore

Library of Congress Cataloging in Publication Data:

Calais, Jean-Louis.
    Quantum chemistry workbook:basic concepts and procedures in the
theory of the electronic structure of matter / by Jean-Louis Calais.
        p.   cm.
    "A Wiley-Interscience publication."
    ISBN 0-471-59435-0
    1. Quantum chemistry.   I. Title.
QD462.C35   1994
541.2'8—dc20   94-2394

Printed in the United States of America

10 9 8 7 6 5 4 3 2 1

CONTENTS

PREFACE

The present book has grown out of two observations:

1. One learns *by doing*, not by reading. This is just as true for a theoretical subject like quantum chemistry as for experimental fields. One of the aims of the book is to get students to *work through it* and in that way discover through their own efforts a number of fundamental and interesting aspects of their field.

2. Even though all the concepts treated in this book are certainly discussed in most textbooks, it is my experience as a teacher for many years that even after having "taken" courses in quantum mechanics, students are lacking both in their understanding of the fundamental concepts and in their ability to handle many of the basic procedures.

I have therefore selected a rather small number of basic concepts and procedures and the point is to help the students to see what these mean in practice for the most important building blocks of matter: free atoms, small molecules, polymers, and crystals. The very comparison of the application of these concepts and procedures to different types of systems is itself of great pedagogical value. It is important that an endeavor of this kind is carefully limited, so that the students do not get lost. The book is consequently limited both in the number of concepts and procedures selected and by staying at an approximation level of one-electron type. This level of approximation is by far the most important one for all applications, and it is at this level that most of the quickly growing number of applications of quantum chemical procedures are carried out.

When it becomes more and more common that chemists of all kinds use quantum chemical procedures and programs as another nearly "experimental" tool, it is imperative that they should know what they are doing. That does not mean that they have to be specialists in quantum chemistry. But they must understand the basic concepts, and that is not at all difficult if they just get a chance to play a little with them. With solid groundwork it is much easier to understand all kinds of instructions for methods and programs.

The aim is also to help students to extract information both from formal treatments and from calculations. It is not at all uncommon that students know what to do to produce certain results while remaining unaware of the wealth of information they are producing. Advanced undergraduate and beginning graduate students in chemistry, physics, and neighboring fields should be able to benefit from the book. In addition, any physicist or chemist who wants to improve her or his basic understanding of quantum mechanics will hopefully find it useful.

It may seem strange that what is said here and in the introduction needs to be said at all. After all, these are just basic requirements for any type of study or research. But it is my experience that for many students their "bad" habits from school have to be changed—they are often far too respectful. If this book can help its users to experience the fun and pleasure that can be associated with playing with concepts and their mathematical expressions, it has fulfilled its purpose.

My own teaching experience teaching has certainly been decisive for the structure of the book. But I would not have had a chance to realize the importance of many of the aspects and perspectives of the book without my experience as a student of a master of teaching—Professor Per-Olov Löwdin. Statements like: "Never accept a statement without checking," "Never be satisfied with a routine derivation—try to find something simpler and more elegant," "Check the consistency," and "Specialize back to known but less general results," to mention a few, are characteristic of his attitude in teaching as well as in his own research. It is that attitude that I hope to spread to readers of this book.

ABOUT THE AUTHOR

JEAN-LOUIS CALAIS is Associate Professor of Quantum Chemistry at the University of Uppsala, Sweden. He is also an Adjunct Professor of Physics at the University of Florida and has served as a senior scientist at the Swedish Natural Sciences Research Council. A leading figure in the field of quantum chemistry, Dr. Calais is an editor of the *International Journal of Quantum Chemistry* and has lectured at numerous research institutions around the world. Dr. Calais received his PhD in quantum chemistry from the University of Uppsala.

I

INTRODUCTION

The main focus of this book is on activating the reader. In particular, Chapters III–VI should, so to speak, not be read—they should be worked through. The book is not supposed to replace textbooks but to complement them and hopefully, to make them more useful. It is quite deliberate that the present book is not "married" to a particular textbook. An essential aspect both of taking a course—learning—and of those walks into the unknown that we call research is to learn to work with a library. Throughout the book references are given (see the references at the end of each chapter) to various textbooks and a few papers in journals, which should be available in libraries associated with physics and chemistry departments. It is not enough to know one textbook. First, their actual contents often differ even though their titles may be similar. And certain things are better expressed in one book than in another. In this connection, "better" usually means better suited to my own particular experiences and temperament.

Most textbooks contain a list of problems at the end of each chapter or at the end of the book. To work through such problems thoroughly is essential for an understanding of the subject, but that is often not enough. One should also *work through the main text*. That requires certain working habits and in a way also a certain familiarity with the subject. One must get used to questioning almost everything and not accepting any statements without careful scrutiny. The present book is intended particularly to help beginning students of quantum mechanics and quantum chemistry to "grow into" such working habits. As a result, they will discover that *one learns by doing and not by reading*.

Even though references are made to certain textbooks of quantum mechanics, the present book should be useful as a complement to almost any textbook or any course in that field. Some background material is obviously necessary. On the other hand, since the book is written with the explicit intention of making the user of the book more active, students are encouraged to start using it directly without awaiting the end of a course. It is very important not to be too respectful when one meets a new subject such as quantum mechanics but to start as soon as possible and have fun. That is another aspect of the present book: to encourage students to use their imagination to invent questions of all kinds, to play with all the formulas that appear, to check the internal consistency of various statements, and so on.

The material exposed in the book is both broader and more restricted than in many other books. The intention is to help the user to build a platform that can serve as a solid base for further applications of quantum mechanics. We have therefore concentrated on some of the most basic properties of such wave functions, which are needed in the theoretical description of the electronic structure of matter. On the other hand, since our primary purpose is to *apply* quantum mechanics, we have chosen a field of applications that is broader than one usually finds in most textbooks. Most textbooks are aimed at students of *either* atomic and molecular physics *or* of solid-state physics. In the present book we study the fundamentals of the theory of *both* atoms and molecules *and* polymers and crystals. In Chapter II we present a minimal, not to say minimalistic, summary of a number of concepts used throughout the book.

There are no explicit answers to the exercises—quite on purpose. In "real life," answers are not available either. One must learn to use the very many possibilities that exist for checking the consistency of a result with data previously available. One of the aims of a course is to become *confident in one's own ability to handle the subject.* To solve problems and find out that one is able to get internally consistent results would seem to be an excellent step toward such a goal. But obviously, the reader should not be left entirely without a helping hand. Nearly all the exercises in this book are therefore constructed such that the answers are available a little later in the text—more or less directly.

Formulas are numbered such that both the chapter and the number are indicated. In references to formulas within the same chapter, only the number is given. If a formula in another chapter is referred to, both the chapter and the number are indicated.

Throughout the book atomic units are used, which means that the mass of the electron, m, and the absolute value of the charge of the electron, $|e|$, are set equal to 1. Planck's constant is set equal to 2π, or expressed in another way, Planck's constant divided by 2π is set equal to 1.

II
SURVEY OF FUNDAMENTALS

In the present chapter we summarize those fundamentals of quantum mechanics that are used directly in subsequent chapters. The central topic of the book is basic properties of wave functions as they appear for some of the most common forms of matter: atoms, small molecules, polymers, and crystals. It is hoped that the reader will get thoroughly familiar with these properties after having worked through the following chapters. The present chapter has a slightly different character than the others, which are filled with exercises.

II.A. BASIC PHILOSOPHY

Like other theories used in science, quantum mechanics can be regarded as a *model*, which agrees with certain aspects of those phenomena we are trying to describe and predict. In our case we are attempting to characterize the electronic structure of matter and to calculate as far as possible the properties of atoms, molecules, and extended systems. Combining quantum mechanics with statistical mechanics, we can then go further and predict the macroscopic properties of matter.

It is presumably not unknown to the reader that quantum mechanics has been and remains surprisingly successful in this task. Both qualitatively and quantitatively this theory lives in a very happy marriage with experimental data. "Qualitatively" refers to the fact that quantum mechanics provides us with an extremely useful conceptual framework, which forms the necessary basis for all thinking and talking about matter at the atomic (electronic) level. On top of that we have the incredible quantitative agreement between theory and experiment in those cases where it has been possible to carry both to extreme precision.

The question of quantitative agreement deserves a special paragraph. The fortunate circumstance that it is possible to solve the Schrödinger equation exactly for the hydrogen atom provided a quick start for quantum mechanics. Further relativistic and quantum electrodynamical corrections have pushed the agreement with experiment to extremely high accuracy (see, e.g., the book by Series referred to at the end of the chapter). In almost all other applications of quantum mechanics to electronic systems, one needs approximate solutions. But approximations can be constructed so as to become better and better systematically. In the case of the hydrogen *molecule*, we certainly do not have an exact solution of the Schrödinger equation. But thanks to intensive mathematical and computational research in the 1950s and the 1960s, we now have an approximate solution of such high accuracy that it has made a new experimental determination of the dissociation energy of H_2 necessary. (See the papers by Kolos–Wolniewicz and Herzberg referred to at the end of the chapter.) This development is obviously not limited to H_2 but extends successively to larger and larger systems and to their interactions.

Our aim in this book is thus to lay a solid conceptual ground for continued work in quantum chemistry. A very important aspect of this is simply to learn ''the craft.'' One does not need to be a specialist in mathematics, but it is essential to master certain mathematical techniques. And the best—probably the only—way to learn is to practice and play with ''illustrations'' of many different types.

II.B. THE SCHRÖDINGER EQUATION

Whenever we want to apply quantum mechanics to a system of elementary particles, the starting point in principle is the time-dependent Schrödinger equation,

$$\mathbf{H}\Psi = \frac{ih}{2\pi}\frac{\partial \Psi}{\partial t}. \tag{II.1}$$

Here $\mathbf{H}$ is the Hamiltonian operator for the system under study—thus defined as soon as we have specified that system. The unknown of the Schrödinger equation is the wave function Ψ, and the principal aim of the book is to help readers become familiar with the basic properties of wave functions for some typical systems.

In subsequent chapters we restrict ourselves to a particular kind of solution of (1)—those for which the probability density $|\Psi|^2 = \Psi^*\Psi$ is independent of time. The states characterized by such wave functions are called *stationary*. They are associated with a specific energy E, and their wave functions depend on time only through an exponential factor:

$$\Psi(\mathbf{r}, t) = \psi(\mathbf{r})e^{-2\pi i E t/h}. \tag{II.2}$$

Substituting (2) in (1), we get the time-independent Schrödinger equation for the function $\psi(\mathbf{r})$,

$$\mathbf{H}\psi(\mathbf{r}) = E\psi(\mathbf{r}). \tag{II.3}$$

Not least, in recent years very interesting progress has been made toward solutions of the time-dependent equation (1). Here we will, however, concentrate on the study of the solutions of the time-independent Schrödinger equation (3).

The physical or chemical system under study—atom, molecule, polymer, crystal, or other—furnishes a Hamiltonian operator of the form

$$\mathbf{H} = \mathbf{T}_N + \mathbf{T}_e + \mathbf{U}_{NN} + \mathbf{U}_{Ne} + \mathbf{U}_{ee}. \tag{II.4}$$

$\mathbf{T}_N$ and $\mathbf{T}_e$ are the kinetic energy operators for the nuclei and electrons, respectively. All particles—nuclei and electrons—interact with Coulombic forces, and these interactions are grouped in nuclear repulsions, $\mathbf{U}_{NN}$, electronic repulsions, $\mathbf{U}_{ee}$, and nuclear–electronic attractions, $\mathbf{U}_{Ne}$.

This Hamiltonian, so to speak, sets the stage for the role the system is going to play. But the physics and chemistry introduced by the Hamiltonian are insufficient for study of a specific situation. A Schrödinger equation (3) with the Hamiltonian (4) has many different types of solutions. We must complement the equation with information about the system under study. Is it completely free? Does it interact in some way with another system? Is it a finite or an extended system? Mathematically, this information takes the form of *boundary conditions* that are imposed on the solutions. Only those solutions that satisfy the boundary conditions imposed are accepted. In the following chapters we study in some detail how different types of boundary conditions lead to very different types of solutions.

A general way of characterizing the role of the boundary conditions is to say that by means of them we select those solutions that are physically and/or chemically reasonable. In many cases this is a perfectly legitimate vocabulary. But we must be aware of the fact that there may exist solutions that do not seem to be acceptable from such points of view, simply because the phenomena described by them have not yet been discovered. Science is always a tentative activity. A set of rules and procedures that have worked well for a long time may have to be modified and extended. An open mind is always necessary to counteract dogmatic temptations.

II.C. POSITION SPACE AND MOMENTUM SPACE

The wave function ψ in the time-independent Schrödinger equation (3) is usually considered to be a function of the spatial variables ("positions") of the particles in the system. In such a case the Hamiltonian must be expressed as an operator working on the position variables. There is nothing sacred about this *position representation*, though. The *momentum representation* is just as important both theoretically and experimentally. One of the aspects to be stressed in this book is to get thoroughly familiar with typical wave functions for the systems under study, in *both* the position and the momentum representations.

The basic transformations connecting wave functions in these two representations are

$$\underline{\phi}(\mathbf{p}) = \frac{1}{\sqrt{8\pi^3}} \int dv \; \phi(\mathbf{r}) e^{-i\mathbf{p}\cdot\mathbf{r}}; \qquad (\text{II.}5a)$$

$$\phi(\mathbf{r}) = \frac{1}{\sqrt{8\pi^3}} \int d\mathbf{p} \; \underline{\phi}(\mathbf{p}) e^{i\mathbf{p}\cdot\mathbf{r}}. \qquad (\text{II.}5b)$$

We use an underlined symbol $\underline{\phi}(\mathbf{p})$ for the counterpart in momentum space of a function $\phi(\mathbf{r})$ in position space. To check the consistency of (5a) and (5b), we need the relation

$$\frac{1}{8\pi^3} \int dv \; e^{i(\mathbf{p}'-\mathbf{p})\cdot\mathbf{r}} = \delta(\mathbf{p}' - \mathbf{p}). \qquad (\text{II.6})$$

Here the right-hand side is Dirac's δ-function, which vanishes everywhere except where its argument itself vanishes. At that point this function—more properly called a *distribution*—is infinite, but in such a way that the integral over it converges and is equal to 1. The integrations in (5) are carried out over all space—position or momentum, as the case may be.

An operator **A** can be expressed in a form specifying how it works in position space. Then we know, for all functions $\phi(\mathbf{r})$ in the domain of the operator, the function

$$\psi(\mathbf{r}) = \mathbf{A}\phi(\mathbf{r}). \tag{II.7}$$

That function has a counterpart in momentum space,

$$\underline{\psi}(\mathbf{p}) = \frac{1}{\sqrt{8\pi^3}} \int dv \; \psi(\mathbf{r}) e^{-i\mathbf{p}\cdot\mathbf{r}}, \tag{II.8}$$

which should be what we get if the operator **A**, specified in momentum space, thus written $\underline{\mathbf{A}}$, works on $\underline{\phi}(\mathbf{p})$:

$$\underline{\psi}(\mathbf{p}) = \underline{\mathbf{A}}\underline{\phi}(\mathbf{p}) = \frac{1}{\sqrt{8\pi^3}} \int dv \; [\mathbf{A}\phi(\mathbf{r})] e^{-i\mathbf{p}\cdot\mathbf{r}}. \tag{II.9}$$

II.D. NORMALIZATION AND ORTHOGONALITY

Following the suggestion by Born, the square of the absolute value of the wave function is interpreted as the probability of finding the system at the "point" specified by the arguments of the wave function. For a one-electron function this means that

$$|\psi(\mathbf{r})|^2 \, dv \tag{II.10}$$

represents the probability of finding the electron in the volume element dv around $\mathbf{r}$. For a two-particle function $\Psi(\mathbf{r}_1, \mathbf{r}_2)$, the quantity

$$|\Psi(\mathbf{r}_1, \mathbf{r}_2)|^2 \, dv_1 \, dv_2 \tag{II.11}$$

represents the probability of finding one particle in dv_1 around $\mathbf{r}_1$ and another in dv_2 around $\mathbf{r}_2$.

This interpretation presupposes that the wave function is normalizable. The particles must be somewhere in space. This implies that the integrals over all space of (10) and (11) must be equal to 1 and 2, respectively. Born's interpretation therefore imposes a general condition on all wave functions: the integral over all space of the square of their absolute value must exist. In such a case the wave function can be multiplied by a suitable constant so as to make the integral equal to the number of particles in the system.

Orthogonality is a related concept. Two wave functions are orthogonal if their overlap integral vanishes:

$$\int dv \; \psi^*(\mathbf{r})\phi(\mathbf{r}) = \mathbf{0}. \tag{II.12}$$

Orthogonality is a concept that appears in many different connections. Eigenfunctions of Hermitian operators associated with different eigenvalues are necessarily orthogonal. A special case of that general theorem is provided by the time-dependent Schrödinger equation (3): wave functions associated with states with different energies are orthogonal. But orthogonality is also a property of sets of functions which simplifies their mathematical handling. In many cases one has the freedom to work with orthogonal basis functions, and that is usually preferable. Several types of orthogonalization procedures have been developed for sets of functions that are not originally orthogonal.

Linear independence is a more general concept than orthogonality. A set of functions ϕ_1, ϕ_2, ϕ_3, . . . , ϕ_n, are linearly independent if the relation

$$\phi_1 c_1 + \phi_2 c_2 + \phi_3 c_3 + \cdots + \phi_n c_n = 0 \tag{II.13}$$

can hold only if *all* the coefficients c_k vanish. For vectors in a three-dimensional Euclidean space, linear independence means that they do not lie in the same plane. Orthogonal vectors, on the other hand, are mutually perpendicular and thus definitely linearly independent.

II.E. TOTAL WAVE FUNCTIONS AND SPIN ORBITALS

Most systems of chemical or physical interest contain more than one electron. One-electron systems, on the other hand, are simpler to handle. In some special one-electron cases we even have exact mathematical solutions of the corresponding Schrödinger equation. The importance of one-electron problems goes far beyond these special cases, though, since they provide us with conceptual, mathematical, and computational tools for the construction of approximate solutions of the Schrödinger equation for many-electron systems.

A one-electron function is also called an *orbital*. That term was coined by Mulliken (see the paper by Coulson referred to at the end of the chapter) to remind us of the fact that in a certain sense one-electron wave functions play a role in quantum mechanics similar to that of *orbits* in classical mechanics. The concept of orbit is meaningless in quantum mechanics because of the uncertainty principle. Thus an orbital is a function of the position of one particle, $\phi(\mathbf{r})$; any coordinate system can obviously be used to represent the vector $\mathbf{r}$.

As we discuss in Section II.I, the electron has a spin—an intrinsic angular momentum. In the present book we are not going to discuss spin-dependent interactions. Still we must take the spin into account, since it influences certain fundamental symmetry properties of many-electron functions. The usual way to handle that—but not the only one—is to endow the electron with a fourth coordinate ζ. The letter x is often used to denote the collection of four coordinates of the electron: three spatial components and one spin component.

$$x = (\mathbf{r}, \zeta). \qquad \text{(II.14)}$$

This is discussed more thoroughly in Section II.I. A function of the four components (14), $\psi(x)$, is called a *spin orbital*. The x in (14) must obviously be carefully distinguished from the Cartesian component x of the vector $\mathbf{r}$. The term *spin orbital* is thus synonymous with the expression *one-electron function*.

A wave function for an N-electron system is a function of the $4N$ (why 4?) variables $x_1, x_2, x_3, \ldots, x_N$:

$$\Psi = \Psi(x_1, x_2, x_3, \ldots, x_N). \qquad \text{(II.15)}$$

It is common practice to use capital letters for many-electron functions and lowercase letters for spin orbitals and orbitals. A many-electron wave function is normalized if

$$\int dx_1 \, dx_2 \, dx_3 \, \cdots \, dx_N \, |\Psi(x_1, x_2, x_3, \ldots, x_N)|^2 = 1,$$

$$\text{(II.16)}$$

9

and two N-electron functions are orthogonal if

$$\int dx_1\, dx_2\, dx_3 \cdots dx_N\, \Psi^*(x_1, x_2, x_3, \ldots, x_N)$$

$$\cdot\, \Phi(x_1, x_2, x_3, \ldots, x_N) = 0. \tag{II.17}$$

We will return to the meaning of integration over spin space in Section II.I.

II.F. MATRIX REPRESENTATIONS OF FUNCTIONS AND OPERATORS

In the present section we review some fundamental concepts of linear algebra that are used over and over again in the theory of the electronic structure of matter. We will often have a reason to expand a known or unknown function in a *basis*—a set of linearly independent functions [cf. (13)] such that any function in the space under consideration can be expanded in them. It is practical to think of the basis as an entity, and for that reason we use a special symbol, a boldface letter, to denote the basis as a whole:

$$\phi = [\phi_1, \phi_2, \phi_3, \ldots, \phi_n]. \tag{II.18}$$

It is important to notice that the basis functions are collected in a *row matrix*.

We expand an arbitrary function f in this basis and denote the coefficients f_i:

$$f = \sum_{j=1}^{n} \phi_j f_j. \tag{II.19}$$

We collect the coefficients of the basis functions in a *column matrix*:

$$\mathbf{f} = \begin{bmatrix} f_1 \\ f_2 \\ f_3 \\ \cdots \\ f_n \end{bmatrix}. \tag{II.20}$$

The expansion (19) can then be regarded as a matrix product:

$$f = \phi \mathbf{f}. \tag{II.21}$$

Such an interpretation apparently presupposes that the basis functions appear in a row and the coefficients in a column.

An operator transforms a function to another function—the function g is obtained when the operator $\mathbf{A}$ works on the function f:

$$g = \mathbf{A} f. \tag{II.22}$$

In particular, the operator can work on one of the basis functions ϕ_i. If the result lies in the space of the basis, it can be expressed in terms of the basis:

$$\mathbf{A}\phi_i = \sum_{j=1}^{n} \phi_j A_{ji}, \quad i = 1, 2, 3, \ldots, n. \tag{II.23}$$

The complete information about the influence of the operator on the n basis functions apparently requires n^2 numbers A_{ji}. These are collected in a square matrix of order n:

$$\mathbf{A} = \begin{bmatrix} A_{11} & A_{12} & A_{13} & \cdots & A_{1n} \\ A_{21} & A_{22} & A_{23} & \cdots & A_{2n} \\ \cdots & \cdots & \cdots & \cdots & \cdots \\ A_{n1} & A_{n2} & A_{n3} & \cdots & A_{nn} \end{bmatrix}. \tag{II.24}$$

We introduce the notation $\mathbf{A}_i$ for the ith column of the matrix $\mathbf{A}$:

$$\mathbf{A}_i = \begin{bmatrix} A_{1i} \\ A_{2i} \\ A_{3i} \\ \cdots \\ A_{ni} \end{bmatrix}. \tag{II.25}$$

Then one of the expansions (23) can be written

$$\mathbf{A}\phi_i = \phi \mathbf{A}_i. \tag{II.26}$$

All the n expansions (23) can be condensed to the expression.

$$\mathbf{A}\phi = \phi\mathbf{A}. \tag{II.27}$$

In words: In the basis ϕ, the operator $\mathbf{A}$ is represented by the matrix $\mathbf{A}$.

Now we have the tools needed to show how the operator works on an arbitrary function f [cf. (22)]:

$$g = \mathbf{A}f = \mathbf{A}\phi\mathbf{f} = \phi\mathbf{Af}. \tag{II.28}$$

On the other hand, we can also write

$$g = \phi\mathbf{g}, \tag{II.29}$$

which means that the components of the function g in the given basis can be written as

$$\mathbf{g} = \mathbf{Af}. \tag{II.30}$$

This is apparently a condensed way of writing the n relations

$$g_j = \sum_{i=1}^{n} A_{ji} f_i, \quad j = 1, 2, 3, \ldots, n. \tag{II.31}$$

There is nothing sacred about the basis ϕ originally introduced. Any other set of n linearly independent functions will do: for example,

$$\chi = [\chi_1, \chi_2, \chi_3, \ldots, \chi_n]. \tag{II.32}$$

Each function in the basis ϕ can be expanded in the basis (32),

$$\phi_i = \chi\alpha_i = \sum_{j=1}^{n} \chi_j \alpha_{ji}, \quad i = 1, 2, 3, \ldots, n, \tag{II.33}$$

and vice versa:

$$\chi_j = \phi\beta_j = \sum_{k=1}^{n} \phi_k \beta_{kj}, \quad j = 1, 2, 3, \ldots, n. \tag{II.34}$$

These two sets of relations can be summarized as

$$\phi = \chi\alpha, \quad \chi = \phi\beta, \tag{II.35}$$

where α and β are $n \times n$ matrices. Consistency apparently requires that

$$\alpha\beta = \beta\alpha = 1. \qquad (\text{II}.36)$$

II.G. SYMMETRY

Symmetry is one of the most fundamental concepts in both classical and quantum mechanics. In the present book we will, however, only touch on such aspects of symmetry as cannot be avoided. This includes the fundamental antisymmetry of many-electron wave functions with respect to permutations of the electronic coordinates, discussed in Section III.B. Atoms are characterized by the fact that they have only one nucleus, and that brings in spherical symmetry with many interesting implications, usually described in terms of angular momentum concepts. Small molecules show other types of symmetry, and certain basic aspects of that are treated in Chapter IV. Polymers have translational symmetry in one dimension and that is treated in some detail in Chapter V. The three-dimensional counterpart of those aspects is the main subject of Chapter VI. Symmetry is such a rich field that it deserves a separate volume. The concepts discussed in the present book center around the most basic aspects: how wave functions transform under certain symmetry operations.

II.H. NORMAL AND HERMITIAN OPERATORS

Those operators that represent physical and chemical quantities must be Hermitian. Other operators will also be needed, however, as auxiliary tools. We therefore review the basic properties of *normal operators*, of which the Hermitian operators constitute a special case. First we need the concept of the *adjoint of an operator*, which we define by the following expression:

$$\langle \Psi | \mathbf{A}^+ | \Phi \rangle = \langle \mathbf{A}\Psi | \Phi \rangle = \langle \Phi | \mathbf{A}\Psi \rangle^*. \qquad (\text{II}.37)$$

This expression also constitutes what is sometimes called the *turnover rule*, which we will use repeatedly in the following chapters.

A Hermitian operator is self-adjoint, $\mathbf{A} = \mathbf{A}^+$, which means that for such an operator (37) reduces to

$$\langle \Psi | \mathbf{A}\Phi \rangle = \langle \Phi | \mathbf{A}\Psi \rangle^*. \qquad \text{(II.38)}$$

It should be noted that in the bra-ket notation the second vertical bar can be suppressed:

$$\langle \Psi | \mathbf{A} | \Phi \rangle = \langle \Psi | \mathbf{A}\Phi \rangle. \qquad \text{(II.39)}$$

A *normal operator* commutes with its adjoint:

$$\mathbf{T}^+\mathbf{T} = \mathbf{T}\mathbf{T}^+. \qquad \text{(II.40)}$$

If the product (40) equals **1**, the operator is apparently unitary.

Two commuting operators have simultaneous eigenfunctions but in general different eigenvalues:

$$\mathbf{T}\phi_k = \lambda_k\phi_k, \qquad \mathbf{T}^+\phi_k = \mu_k\phi_k. \qquad \text{(II.41)}$$

Assuming that the functions ϕ_k are normalized, we get, using (37),

$$\lambda_k = \langle \phi_k | \mathbf{T}\phi_k \rangle = \langle \mathbf{T}^+\phi_k | \phi_k \rangle = \mu_k^*. \qquad \text{(II.42)}$$

Two eigenfunctions ϕ_k and ϕ_l associated with different eigenvalues $\lambda_k \neq \lambda_l$ are orthogonal, which can be seen, for example, by multiplying (41) by ϕ_l^*, integrating, and using (37):

$$\langle \phi_l | \mathbf{T}\phi_k \rangle = \lambda_k\langle \phi_l | \phi_k \rangle; \qquad \text{(II.43}a\text{)}$$

$$\langle \phi_l | \mathbf{T}\phi_k \rangle = \langle \mathbf{T}^+\phi_l | \phi_k \rangle = \lambda_l\langle \phi_l | \phi_k \rangle; \qquad \text{(II.43}b\text{)}$$

$$(\lambda_l - \lambda_k)\langle \phi_l | \phi_k \rangle = 0. \qquad \text{(11.43}c\text{)}$$

In the special case when $\mathbf{T}^+ = \mathbf{T}$, (42) implies that $\lambda_k^* = \lambda_k$ (i.e., Hermitian operators have real eigenvalues).

II.I. SPIN

Electrons are fermions with spin $\frac{1}{2}$. That statement has deep roots in angular momentum theory, which we do not discuss in the present book. Here we review only the formalism connected with spin coordinates and spin functions.

Spin space is a very particular kind of space. The *spin coordinate* can take only two possible *discrete* values, which we call $+1$ and -1. Thus the spin coordinate ζ can be equal to ± 1—no other values exist. We will work with two *spin functions*, traditionally denoted $\alpha(\zeta)$ and $\beta(\zeta)$. They are completely defined if their values for $\zeta = \pm 1$ are given. The actual numerical values are immaterial, but for normalization purposes one usually chooses the following expressions:

$$\alpha(+1) = 1; \quad \beta(+1) = 0;$$

$$\alpha(-1) = 0; \quad \beta(-1) = 1. \tag{II.44}$$

One speaks of "integration over spin space," which, strictly speaking, is a misnomer for "summation over spin space." Given a function $f(\zeta)$ of the spin coordinate, we thus have

$$\int d\zeta\, f(\zeta) = f(+1) + f(-1). \tag{II.45}$$

The primary reason for this nomenclature is that summation over spin space very often occurs together with integration over "ordinary space."

Using (44) and (45), we see directly that the two spin functions are orthogonal,

$$\int d\zeta\, \alpha(\zeta)\beta(\zeta) = \alpha(+1)\beta(+1) + \alpha(-1)\beta(-1) = 0, \tag{II.46}$$

and normalized,

$$\int d\zeta\, \alpha(\zeta)\alpha(\zeta) = \int d\zeta\, \beta(\zeta)\beta(\zeta) = 1. \tag{II.47}$$

A *spin orbital* is any function of the combined variable $x = (\mathbf{r}, \zeta)$. Since spin space is spanned by the two functions (44) (in other words, α and β form a complete set in spin space), any spin orbital can be expanded in these functions:

$$\psi(x) = \psi_+(\mathbf{r})\alpha(\zeta) + \psi_-(\mathbf{r})\beta(\zeta). \tag{II.48}$$

The functions $\psi_\pm(\mathbf{r})$ are then functions of the spatial variable $\mathbf{r}$. The expression (48) is the most general form of a spin orbital. If we write the basis functions as a row vector [cf. (18)] and the coefficients $\psi_{\pm}(\mathbf{r})$ is a column vector, we can write (48) as a matrix product:

$$\psi(x) = [\alpha(\zeta) \quad \beta(\zeta)] \begin{bmatrix} \psi_+(\mathbf{r}) \\ \psi_-(\mathbf{r}) \end{bmatrix}. \tag{II.49}$$

The column with the orbital components of the spin orbital,

$$\boldsymbol{\psi}(\mathbf{r}) = \begin{bmatrix} \psi_+(\mathbf{r}) \\ \psi_-(\mathbf{r}) \end{bmatrix}, \tag{II.50}$$

is called a *spinor*. In the remaining chapters of the present book we are going to use only two special cases of (49):

$$\psi(\mathbf{r})\alpha(\zeta) = [\alpha(\zeta) \quad \beta(\zeta)] \begin{bmatrix} \psi(\mathbf{r}) \\ 0 \end{bmatrix}; \tag{II.51a}$$

$$\psi(\mathbf{r})\beta(\zeta) = [\alpha(\zeta) \quad \beta(\zeta)] \begin{bmatrix} 0 \\ \psi(\mathbf{r}) \end{bmatrix}. \tag{II.51b}$$

In other words, the spin orbitals will be simple products of an orbital and a spin function. In more elaborate treatments, general spin orbitals like (48) are used.

II.J. ONE-ELECTRON APPROXIMATIONS

In principle we start our applications of quantum mechanics by setting up the time-independent Schrödinger equation for the N-electron system under study.

$$\mathbf{H}\Psi(x_1, x_2, x_3, \ldots, x_N) = E\Psi(x_1, x_2, x_3, \ldots, x_N).$$

$$\tag{II.52}$$

Only those solutions of (52) are admitted which satisfy the boundary conditions that have been imposed, which should correspond to the particular chemical or physical situation of interest. To formulate those boundary conditions may actually turn out to be one of the more difficult parts of the problem.

To solve the N-electron equation (52) is a formidable problem, and at least so far it has only been possible to construct approximate solutions, which can, however, be quite accurate when N is not too large. The probably most important approximation amounts to replacing (52) by a *one-electron approximation*, a set of coupled equations for spin orbitals. One way toward that goal first replaces the many-electron wave function Ψ in (52) by a special kind of many-electron function that is more easily handled: a single determinant of one-electron functions,

$$\Psi(x_1, x_2, x_3, \ldots, x_N) \sim D(x_1, x_2, x_3, \ldots, x_N)$$

$$= \frac{1}{\sqrt{N!}} \begin{vmatrix} \psi_a(x_1) & \psi_b(x_1) & \cdots & \psi_{k_N}(x_1) \\ \psi_a(x_2) & \psi_b(x_2) & \cdots & \psi_{k_N}(x_2) \\ \cdots & \cdots & \cdots & \cdots \\ \psi_a(x_N) & \psi_b(x_N) & \cdots & \psi_{k_N}(x_N) \end{vmatrix}. \quad \text{(II.53)}$$

This is a rather drastic approximation, since one can show that any antisymmetric (more about that in Section III.B) N-electron function, thus also Ψ in (52), can be written as an *infinite sum* of determinants. But the great merit of (53) is that we can replace the unwieldy N-electron problem (52) by a set of coupled one-electron equations for the so-far unknown spin orbitals in (53),

$$\mathbf{h}_{\text{eff}}(1)\psi_k(x_1) = \epsilon_k \psi_k(x_1). \quad \text{(II.54)}$$

How that is done is beyond the scope of this book. The Hartree–Fock equations (54) can then be solved, even though that is by no means a trivial task.

Another road toward one-electron approximations is based on the Hohenberg–Kohn theorem, which states that the total energy of any N-electron system is a unique functional of the electron density. That theorem forms the basis of density functional theory, which has been extremely successful, particularly for extended systems, where N is very large.

REFERENCES

Atkins, P. W., *Molecular Quantum Mechanics*, Oxford University Press, Oxford, 1970; second edition, 1983.

Bransden, B. H., and C. J. Joachain, *Introduction to Quantum Mechanics*, Longman Scientific and Technical; co-published with John Wiley & Sons, Inc.; Harlow, Essex, England, and New York, 1989.

Cohen-Tannoudji, C., B. Liu, and F. Laboë, *Mécanique quantique*, Hermann, Paris, 1973; second edition 1977; English translation *Quantum Mechanics*, Wiley, New York, 1977.

Coulson, C. A., in *Molecular Orbitals in Chemistry, Physics, and Biology*, ed. P.-O. Löwdin and B. Pullman, Academic Press, New York, 1969, p. 4.

Herzberg, G., *J. Mol. Spectrosc.* **33,** 147 (1970).

Kolos, W., and L. Wolniewicz, *J. Chem. Phys.* **49,** 404 (1968).

Landau, L. D., and E. M. Lifshitz, *Quantum Mechanics*, Pergamon Press, Oxford, 1958.

Messiah, A., *Mécanique quantique*, Dunod, Paris, 1959; English translation *Quantum Mechanics*, Wiley, New York, 1961.

Pauling, L., and E. B. Wilson, *Introduction to Quantum Mechanics*, McGraw-Hill, New York, 1935.

Series, G. W., *Spectrum of Atomic Hydrogen*, Oxford University, Oxford, 1957.

III
ATOMS

Atoms and ions constitute the simplest electronic systems. Their primary characteristic is obviously the fact that atoms and ions have only one nucleus. In the present chapter we examine such systems from the various points of view introduced in Chapter II. We recall that in the present volume we consider only a level of approximation where we have as many spin orbitals as we have electrons. The total wave function is then a single determinant. We start out with the hydrogen atom, where we know the exact solution. Even if many electron atoms have a number of complicating features, the hydrogen atom still forms a *paradigm* for them also.

III.A. THE HYDROGEN ATOM

Basic Characteristics of the Solutions

As can be seen in any textbook of basic quantum mechanics, the bound-state solutions of the Schrödinger equation for the system consisting of one electron and a nucleus with charge Z can be written

$$\psi_{nlm}(\mathbf{r}) = R_{nl}(r)Y_{lm}(\vartheta, \varphi), \qquad \text{(III.1)}$$

where both the radial functions $R_{nl}(r)$ and the spherical harmonics $Y_{lm}(\vartheta, \varphi)$ are well-known functions. An important aspect of the solutions $\psi_{nlm}(\mathbf{r})$ of the hydrogen atom problem is just the fact that they are factorized in a radial and an angular function.

1. Which property of the Hamiltonian for the hydrogen atom is responsible for this factorization?

2. Which values are possible for the three quantum numbers n, l, and m?

3. In which intervals are the three polar coordinates r, ϑ, and φ defined?

The properties of the functions in (1) are related to the fact that certain boundary conditions have been imposed on the solutions. For the angular functions the property of single valuedness plays an important role.

EXERCISES

4. Look up the explicit expressions for the spherical harmonics with $l = 1$. How many such functions are there? What does the expression *one-valued* mean for these functions with respect so the two variables? Is the function $\cos(\varphi/2)$ one-valued on the sphere?

5. Plot the functions **(a)** $Y_{10}(\vartheta, \varphi)$ and **(b)** $Y_{11}(\vartheta, \pi/2)$ as functions of ϑ.

The boundary conditions for the radial functions are of a different nature, which is related to the values that are possible for their argument r.

6. Look up the first few radial solutions of the hydrogen atom problem. In which interval of r are we interested? How do they behave for large r? How do they before for small r? What type of boundary conditions have been imposed?

7. Plot the hydrogenic radial functions **(a)** $R_{10}(r)$; **(b)** $R_{20}(r)$; and **(c)** $R_{21}(r)$ as functions of r.

It is very important to get a good feeling for the behavior of the solutions of the time-independent Schrödinger equation for the hydrogenlike ions, since these or related functions are used in practically all quantum mechanical calculations.

8. Look up the radial functions for the ground state of the ions H, Li^{2+}, B^{4+}, and O^{7+}. Plot them as functions of r in the same diagram. What is common and what is different between these functions?

9. Which values do these functions have for $r = 0$? Interpretation?

It is instructive to identify in some detail what we mean by symmetry for the hydrogen atom orbitals. Normally, statements of that kind refer to the transformation properties of the functions under certain geometric operations. For a system such as the hydrogen atom, those geometric operations constitute the full three-dimensional rotation group. For any such rotation characterized by the Euler angles α, β, and γ we can specify explicitly how the spherical harmonics transform. Here we consider only some special cases.

EXERCISES

10. What happens to an s-function under a rotation? Does the result depend on the choice of rotation axis?

11. What happens to a p_z-function under a rotation by $\pi/2$, π, and $3\pi/2$ around the x-axis?

12. What happens to a p_x-function under reflection in the xz-plane and under reflection in the yz-plane?

13. Can an s-function be transformed to a p-function, or vice versa? Why or why not?

The spherical harmonics $Y_{lm}(\vartheta, \varphi)$ are perhaps the most widely used functions in quantum mechanics. They are well known and their properties have been investigated in great detail. It is essential to become thoroughly familiar with these functions.

14. How many spherical harmonics are there for a given value of l?

15. Which values of l are possible? Make a list of all spherical harmonics for $l = 0$, 1, and 2. They are usually given in polar coordinates.

16. Use the transformation from polar to Cartesian coordinates to express the spherical harmonics for $l = 1$ in Cartesian coordinates. Take a careful look at these functions in these two representations. Which one shows most clearly how they transform under rotations?

The spherical harmonics are orthogonal with respect to both their arguments. The standard notation $Y_{lm}(\vartheta, \varphi)$ refers to normalized functions. The general expression for these important properties—their orthonormality—is

$$\int_0^\pi \sin \vartheta \, d\vartheta \int_0^{2\pi} d\varphi \, Y_{l'm'}^*(\vartheta, \varphi) Y_{lm}(\vartheta, \varphi) = \delta_{l'l}\delta_{m'm}. \quad \text{(III.2)}$$

17. Illustrate (2) by calculating the integrals explicitly for (a) $l' = 0$, $m' = 0$, $l = 1$, $m = 0$; (b) $l' = 0$, $m' = 0$, $l = 1$, $m = 1$; (c) $l' = 0$, $m' = 0$, $l = 0$, $m = 0$; (d) $l' = 1$, $m' = 0$, $l = 1$,

$m = -1$; and (e) $l' = 1$, $m' = 0$, $l = 1$, $m = 0$. Look up the explicit expressions for the functions needed in a suitable textbook. Try to identify which particular properties of the functions cause the orthogonality in each case.

The spherical harmonics show how the wave functions depend on the angular variables. Once we have found that these functions constitute exact solutions of the angular problem, it remains to set up and solve the radial equation.

EXERCISE

18. The radial Hamiltonian for the hydrogen atom problem consists of a kinetic and a potential energy term. How do they look? The potential is an *effective potential* in the sense that it consists of the primary Coulomb term and a term due to the angular part of the problem. Write up that last centrifugal term and plot it as a function of r for $l = 0$, 1, and 2 in the same diagram. Then plot in another diagram the corresponding three full effective potentials. What do these plots tell us?

Energy Levels and Other Properties

The time-independent Schrödinger equation for the hydrogen atom problem has basically two kinds of solutions, differing in the sign of the energy eigenvalue. In a state with negative energy the electron and the proton form a *bound state*. If, on the other hand, the energy eigenvalue is positive, the electron is no longer bound to the proton even though the two particles interact. All positive values of the energy are compatible with the boundary conditions: the positive eigenvalues form a continuous spectrum. Negative eigenvalues, on the other hand, are compatible with the boundary conditions only for a discrete (but infinite) set of energies. We will not discuss the solutions corresponding to positive energies here; they can be found, for example, in Landau–Lifshitz (see references at the end of this chapter). The negative energies E_n are thus the eigenvalues of the equation

$$\mathbf{H}\psi_{nlm}(\mathbf{r}) = E_n\psi_{nlm}(\mathbf{r}), \qquad E_n < 0. \qquad \text{(III.3)}$$

We notice already the surprising fact that although the wave functions depend on three quantum numbers n, l, and m, the energy depends only on n. We will return to this aspect later.

EXERCISE

19. Look up the formula for the energy eigenvalues. Plot these levels as a function of n, for $n = 1, 2, 3, \ldots$. Show in the diagram energy scales in **(a)** atomic units obtained when $e = m = 1$ and $h = 2\pi$ (hartrees); **(b)** electron volts. How many electron volts are needed to ionize the hydrogen atom in a state with (a) $n = 1$, (b) $n = 3$; and (c) $n = 35$?

Physical and chemical properties of microscopic systems are associated with operators. To any property that can be measured corresponds a Hermitian operator (what is that, and why do we need the Hermiticity?). The value in a state characterized by the wave function ψ of the property represented by the operator $\mathbf{A}$ is given by the expectation (average) value,

$$\langle \mathbf{A} \rangle = \frac{\langle \psi | \mathbf{A} | \psi \rangle}{\langle \psi | \psi \rangle}; \qquad \text{(III.4}a\text{)}$$

$$\langle \psi | \mathbf{A} | \psi \rangle = \int dv \, \psi^*\mathbf{A}\psi; \qquad \text{(III.4}b\text{)}$$

$$\langle \psi | \psi \rangle = \int dv \, \psi^*\psi. \qquad \text{(III.4}c\text{)}$$

---

20. Calculate for hydrogenlike ions the average distance between the electron and the nucleus as a function of the nuclear charge Z in states with the following quantum numbers: **(a)** $n = 1$, $l = m = 0$; **(b)** $n = 2$, $l = m = 0$; and **(c)** $n = 2$, $l = 1$, $m = 0$. The operator **A** in this case is just the number r.

21. Calculate the average momentum for the same states. The operator **A** is then $\mathbf{p} = -(ih/2\pi)\nabla$.

22. Calculate the average kinetic energy for the same states. The operator **A** is then $-(h^2/8\pi^2 m)\Delta$. Compare the results with the total energy in these states.

Momentum-Space Wave Functions

So far we have worked with wave functions in position space. An important characteristic of quantum mechanics is, however, the possibility of using different *representations* of wave functions and operators. Given a function in position space, $\phi(\mathbf{r})$, we can construct its counterpart in momentum space by means of a Fourier transform:

$$\underline{\phi}(\mathbf{p}) = \frac{1}{\sqrt{8\pi^3}} \int dv \, \phi(\mathbf{r}) e^{-i\mathbf{p}\cdot\mathbf{r}}. \qquad \text{(III.5)}$$

Since (for a summary of the properties of the Dirac δ-function, see Messiah, Appendix A)

$$\int d\mathbf{p} \, e^{i\mathbf{p}\cdot(\mathbf{r}-\mathbf{r}')} = 8\pi^3 \delta(\mathbf{r} - \mathbf{r}'), \qquad \text{(III.6)}$$

we can easily "invert" (5):

$$\frac{1}{\sqrt{8\pi^3}} \int d\mathbf{p}\, \underline{\phi}(\mathbf{p}) e^{i\mathbf{p}\cdot\mathbf{r}} = \frac{1}{8\pi^3} \int dv'\, \phi(\mathbf{r}') \int d\mathbf{p}\, e^{i\mathbf{p}\cdot(r-r')}$$

$$= \int dv'\, \phi(\mathbf{r}')\delta(\mathbf{r} - \mathbf{r}') = \phi(\mathbf{r}).$$

(III.7)

Notice that all the integrations in (5), (6), and (7) are carried over all space (position or momentum, as the case may be). The two functions $\phi(\mathbf{r})$ and $\underline{\phi}(\mathbf{p})$ are in general different functions of their respective arguments. They are, however, closely related—they are each other's *counterparts*—which is why one often chooses a notation to stress that fact. In the present text, momentum space functions are underlined.

EXERCISES

23. What can be said about the counterpart in momentum space of a function $f(\mathbf{r})$ in position space, which is independent of the angles ϑ and φ?

24. Show that, given (5), the counterpart of the function $\phi^*(\mathbf{r})$ is $\underline{\phi}^*(-\mathbf{p})$.

25. What can be said about the momentum-space counterpart of a real function in position space?

For historical reasons most of quantum mechanics "happens" in position space, in the sense that operators and wave functions are usually represented in that space. Nothing prevents us from working in momentum space, however, and it is very instructive to calculate explicit expressions in momentum space.

EXERCISE

26. Calculate the momentum-space functions $\underline{R}_{1s}(p)$ and $\underline{R}_{2s}(p)$ for the $1s$ and $2s$ states of the hydrogen atom. Plot them.

An important tool for going between momentum and position space (as well as for many other manipulations) is the expansion of a plane wave in spherical harmonics (see any of the books in the references at the end of Chapter II):

$$e^{i\mathbf{p}\cdot\mathbf{r}} = 4\pi \sum_{l=0}^{\infty} \sum_{m=-l}^{l} i^l j_l(pr) Y_{lm}^*(\vartheta_p, \varphi_p) Y_{lm}(\vartheta, \varphi). \quad \text{(III.8)}$$

Here $(p, \vartheta_p, \varphi_p)$ are the polar coordinates of the vector $\mathbf{p}$, and (r, ϑ, φ) are those of $\mathbf{r}$. The function $j_l(pr)$ is the lth-order spherical Bessel function. Explicit expressions for these functions and a summary of their most important properties can be found at the end of Messiah I, Appendix B2.

EXERCISES

27. The plane wave $e^{i\mathbf{p}\cdot\mathbf{r}}$ is characterized by the two vectors $\mathbf{p}$ and $\mathbf{r}$. In what way do their directions show up on the right-hand side of (8)?

28. Where do the lengths of these vectors come in on the right-hand side? On the left-hand side, the two vectors **p** and **r** appear symmetrically. Is that true also on the right-hand side?

29. Look up the first two spherical Bessel functions, $j_0(x)$ and $j_1(x)$, and plot them.

The spherical Bessel functions are nearly as ubiquitous as the spherical harmonics, and this is a good place to start becoming familiar with them. The expansion (8) can be seen as an expansion of the plane wave, considered as a function of the polar angles, in the complete set of functions of these variables that is constituted by the spherical harmonics. In other words, since the spherical harmonics form a complete set of functions of the angles, any ("reasonable") function of these angles can be expanded in them. In the particular case of a plane wave it is also possible to find the coefficients explicitly.

EXERCISES

30. Write out explicitly the first few terms of (8), for $l = 0$ and $l = 1$, together with the associated values of m. Look up (e.g., in Messiah I) the explicit expressions for the corresponding spherical Bessel functions and spherical harmonics.

31. The real part of the plane wave in (8) is $\cos(\mathbf{p} \cdot \mathbf{r})$. Identify the real part of the expansion on the right-hand side of (8) as a way of getting an expansion of $\cos(\mathbf{p} \cdot \mathbf{r})$ in spherical harmonics. Carry out the same procedure for the imaginary part of (8).

We use (8) in (5):

$$\underline{\phi}(\mathbf{p}) = \frac{1}{\sqrt{8\pi^3}} \int dv \, \phi(\mathbf{r}) e^{i\mathbf{p}\cdot\mathbf{r}}$$

$$= \frac{4\pi}{\sqrt{8\pi^3}} \sum_{l=0}^{\infty} \sum_{m=-l}^{l} (-i)^l \int dv \, \phi(\mathbf{r}) j_l(pr)$$

$$\cdot Y_{lm}(\vartheta_p, \varphi_p) Y_{lm}^*(\vartheta, \varphi). \qquad (III.9)$$

If the function $\phi(\mathbf{r})$ is of the form $R_{nl'}(r) Y_{l'm'}(\vartheta, \varphi)$ [i.e., if it is an *atomic orbital* (AO)] we can begin by carrying out the angular integration in (9), using (2). The double sum in (8) is then reduced to a single term, and the momentum-space counterpart of $R_{nl}(r) Y_{lm}(\vartheta, \varphi)$ becomes

$$\underline{\phi}(\mathbf{p}) = \frac{4\pi}{\sqrt{8\pi^3}} (-i)^{l'} Y_{l'm'}(\vartheta_p, \varphi_p) \int_0^\infty r^2 \, dr \, R_{n'l'}(\mathbf{r}) j_{l'}(pr).$$

$$(III.10)$$

Notice that this function is also of the form (1). [Questions: Which is the radial part $R_{n'l'}(p)$ of the momentum-space function?] And not only in the sense that both orbitals are factorized in a radial and an angular part: the angular factors are the same in a certain sense—which one?

EXERCISE

32. Calculate explicitly and plot the radial part $R_{2p}(p)$ of the momentum-space counterpart of the hydrogenic $2p$ orbital.

Momentum-Space Expectation Values

Even though it may sound counterintuitive, the information content is exactly the same whether we work in position space or in momentum space. To see that more explicitly, we rewrite a typical expectation value in terms of momentum-space wave functions. Using (7), we can write the normalization constant (4c) as

$$\langle \psi | \psi \rangle = \int dv \ \psi^*(\mathbf{r})\psi(\mathbf{r})$$

$$= \frac{1}{8\pi^3} \int d\mathbf{p}' \, d\mathbf{p} \ \underline{\psi}^*(\mathbf{p}')\underline{\psi}(\mathbf{p}) \int dv \ e^{i(\mathbf{p} - \mathbf{p}')\cdot\mathbf{r}}$$

$$= \int d\mathbf{p}' \, d\mathbf{p} \ \underline{\psi}^*(\mathbf{p}')\underline{\psi}(\mathbf{p})\delta(\mathbf{p} - \mathbf{p}')$$

$$= \int d\mathbf{p} \ \underline{\psi}^*(\mathbf{p})\underline{\psi}(\mathbf{p}) = \langle \underline{\psi} | \underline{\psi} \rangle. \tag{III.11}$$

With an operator $\mathbf{A}(\mathbf{r})$ working in position space between the functions as in (4b), we get, similarly,

$$\langle \psi | \mathbf{A} | \psi \rangle = \int dv \ \psi^*(\mathbf{r})\mathbf{A}(\mathbf{r})\psi(\mathbf{r})$$

$$= \frac{1}{8\pi^3} \int d\mathbf{p}' \, d\mathbf{p} \ \underline{\psi}^*(\mathbf{p}')\underline{\psi}(\mathbf{p}) \int dv \ \mathbf{A}(\mathbf{r})e^{i(\mathbf{p} - \mathbf{p}')\cdot\mathbf{r}}$$

$$= \int d\mathbf{p}' \, d\mathbf{p} \ \underline{\psi}^*(\mathbf{p}')\underline{\mathbf{A}}(\mathbf{p} - \mathbf{p}')\underline{\psi}(\mathbf{p}). \tag{III.12}$$

Depending on the character of the operator, a certain care is needed. Here we limit ourselves to a multiplicative operator which multiplies only the function of $\mathbf{r}$ on which it is working. In such a case we can use its Fourier transform:

$$\underline{\mathbf{A}}(\mathbf{p}) = \frac{1}{8\pi^3} \int dv \ \mathbf{A}(\mathbf{r})e^{i\mathbf{p}\cdot\mathbf{r}}. \tag{III.13}$$

EXERCISES

33. Which is the counterpart in momentum space to the identity operator in position space?

34. "Invert" (13) and find the operator in position space that corresponds to the identity operator in momentum space.

35. Construct an operator in momentum space whose counterpart in position space depends only on the distance r.

36. Which is the counterpart in momentum space of a screened Coulomb potential $e^{-\lambda r}/r$?

37. Use the result of Exercise 36 to show that the momentum-space counterpart of an unscreened Coulomb potential is $4\pi/p^2$.

Schrödinger Equation in Momentum Space

To transform an equation in position space like

$$\mathbf{H}\psi(\mathbf{r}) = E\psi(\mathbf{r}), \qquad (\text{III}.14)$$

we multiply both sides of $(8\pi)^{-3/2}e^{-i\mathbf{p}\cdot\mathbf{r}}$ and integrate over $\mathbf{r}$. On the right-hand side we then get $E\underline{\psi}(\mathbf{p})$. To the left we first need the term corresponding to the kinetic energy:

$$-\frac{1}{2}\Delta\psi(\mathbf{r}) = -\frac{1}{2}\cdot\frac{1}{\sqrt{8\pi^3}}\int d\mathbf{p}'\,\underline{\psi}(\mathbf{p}')\,\Delta e^{i\mathbf{p}'\cdot\mathbf{r}}$$

$$= \frac{1}{2}\cdot\frac{1}{\sqrt{8\pi^3}}\int d\mathbf{p}'\,\underline{\psi}(\mathbf{p}')p'^2 e^{i\mathbf{p}'\cdot\mathbf{r}}; \qquad (\text{III}.15a)$$

$$\frac{1}{\sqrt{8\pi^3}} \int dv \left[-\frac{1}{2} \Delta\psi(\mathbf{r}) \right] e^{i\mathbf{p}\cdot\mathbf{r}} = \frac{p^2}{2} \underline{\psi}(\mathbf{p}). \quad \text{(III.15b)}$$

EXERCISE

38. Carry through the last step in (15*b*) in detail. Notice how the derivative operator in position space corresponds to a multiplicative operator in momentum space.

For a potential energy operator that is multiplicative in position space, we get

$$\frac{1}{\sqrt{8\pi^3}} \int dv \, V(\mathbf{r})\psi(\mathbf{r}) e^{-i\mathbf{p}\cdot\mathbf{r}} = \int d\mathbf{p}' \, \underline{V}(\mathbf{p} - \mathbf{p}')\underline{\psi}(\mathbf{p}').$$

$$\text{(III.16)}$$

EXERCISE

39. Use the result of Exercise 37 to calculate (16) when $V(\mathbf{r}) = 1/r$.

Densities and Form Factors for One-Electron Systems

Electron Density
The electron density $\rho(\mathbf{r})$ of a system in a state described by the wave function $\psi(\mathbf{r})$ is given by the formula

$$\rho(\mathbf{r}) = \psi(\mathbf{r})^*\psi(\mathbf{r}). \qquad \text{(III.17)}$$

33

40. In what way is the requirement that the wave function be normalized to 1 related to the definition (17)?

41. The density is a physical (or chemical!) quantity, and as such it must be real. How can we be sure that the definition (17) guarantees that?

The properties of the wave function and those of the corresponding density are related but not necessarily in a trivial way. The density obviously vanishes where the wave function does. For an atom or ion the density tends to zero far from the nucleus just as the wave function must do. The rate of that decay is also of great importance.

EXERCISES

42. Calculate and plot the electron density of the hydrogen atom in its ground state as a function of the distance from the nucleus.

43. Calculate and plot the electron density of He^+ and Li^{2+} in their ground states. Compare the decay rates as r increases, for the three systems H, He^+, and Li^{2+}.

44. Calculate and plot the radial electron density of the hydrogen

atom in one of its $2p$ states. How do the densities in the states $2p_x$, $2p_y$, and $2p_z$ differ from each other?

Momentum Distribution

The density in momentum space is usually called the momentum distribution. It is obtained from the wave function in momentum space in the same way as the corresponding quantities in position space:

$$\underline{\rho}(\mathbf{p}) = \underline{\psi}(\mathbf{p})^*\underline{\psi}(\mathbf{p}). \qquad \text{(III.18)}$$

It is very instructive to relate the momentum distribution to the wave function in position space by means of (5):

$$\underline{\rho}(\mathbf{p}) = \frac{1}{8\pi^3} \int dv' \ \psi^*(\mathbf{r}')e^{i\mathbf{p}\cdot\mathbf{r}'} \int dv \ \psi(\mathbf{r})e^{-i\mathbf{p}\cdot\mathbf{r}}$$

$$= \frac{1}{8\pi^3} \int dv' \ dv \ \psi^*(\mathbf{r}')\psi(\mathbf{r})e^{i\mathbf{p}\cdot(\mathbf{r}'-\mathbf{r})}. \qquad \text{(III.19)}$$

Notice that in the integrand we have a product like (17) but *with different arguments* $\mathbf{r}$ and $\mathbf{r}'$ in the two functions.

EXERCISES

45. Express the density in position space in terms of the momentum-space function.

46. Calculate and plot the momentum distribution for the $1s$ and $2s$ states of the hydrogen atom.

Form Factor

The Fourier transform of the density is called the form factor:

$$F(\mathbf{p}) = \int dv \, \rho(\mathbf{r}) e^{i\mathbf{p}\cdot\mathbf{r}}. \qquad \text{(III.20)}$$

We multiply by $e^{-i\mathbf{p}\cdot\mathbf{r}'}$ and integrate over $\mathbf{p}$, using (6):

$$\int d\mathbf{p} \, F(\mathbf{p}) e^{-i\mathbf{p}\cdot\mathbf{r}'} = \int dv \, \rho(\mathbf{r}) \int d\mathbf{p} \, e^{i\mathbf{p}\cdot(\mathbf{r}-\mathbf{r}')}$$

$$= 8\pi^3 \int dv \, \rho(\mathbf{r})\delta(\mathbf{r} - \mathbf{r}') = 8\pi^3 \rho(\mathbf{r}').$$

$$\text{(III.21)}$$

Thus the form factors are the coefficients in the plane-wave expansion of the density.

EXERCISES

47. What can be said about the form factors corresponding to a density that is isotropic (i.e., depends only on the distance r)?

48. Calculate the form factor in the $1s$ state of the hydrogen atom.

49. Calculate the form factor in the $2p$ state of the hydrogen atom.

Reciprocal Form Factors

We can also make a Fourier transform of the momentum distribution (18):

$$B(\mathbf{r}) = \int d\mathbf{p} \, \underline{\rho}(\mathbf{p}) e^{-i\mathbf{p}\cdot\mathbf{r}}. \qquad\qquad \text{(III.22)}$$

That function is called the reciprocal form factor, and it is apparently a function in position space.

EXERCISES

50. Calculate the reciprocal form factor in the ground state of the hydrogen atom.

51. Express the reciprocal form factor in terms of the position-space wave function.

III.B. ATOMS WITH MORE THAN ONE ELECTRON

Antisymmetry

Even though a many-electron atom is a considerably more complicated system than a hydrogen atom, there are a number of aspects that can be taken over more or less directly from the one-electron case. That is particularly true in the Hartree–Fock approximation, which tries to simulate a kind of one-electron behavior. The electrons in a system consisting of N electrons and a nucleus with charge Z (atomic units) are described by an N-electron wave function $\Psi(x_1, x_2, x_3, \ldots, x_N)$ satisfying the time-independent Schrödinger equation

$$\mathbf{H}\Psi(x_1, x_2, x_3, \ldots, x_N) = E\Psi(x_1, x_2, x_3, \ldots, x_N).$$

$$\text{(III.23)}$$

The coordinate x_i stands for the *four* components $(\mathbf{r}_i, \zeta_i)$, where ζ_i is the spin coordinate of electron i and $\mathbf{r}_i$ represents its three "ordinary" spatial coordinates. At this stage of the theory we need the spin coordinates "only" to get the right type of symmetry with respect to permutations of the electronic coordinates for the total wave function. Since the electrons are fermions, all wave functions for systems of electrons must be *antisymmetric under all permutations of the electronic coordinates.* That is an absolutely fundamental requirement and it is fortunately easy to satisfy it. More explicitly, this means that

$$\mathbf{P}\Psi(x_1, x_2, x_3, \ldots, x_N) = (-1)^p \Psi(x_1, x_2, x_3, \ldots, x_N).$$

$$(\text{III}.24)$$

Here $\mathbf{P}$ is a permutation of the N variables x_i and p is the parity of that permutation. There are $N!$ possible permutations of N objects and any one of them can be written as a product of transpositions (i.e., permutations involving only two objects). The number of such transportation is a characteristic property of the permutation and that number is its parity. A transposition like $\mathbf{P}_{12}$ which interchanges x_1 and x_2 thus has parity 1, and we have for a function that is properly antisymmetric

$$\mathbf{P}_{12}\Psi(x_1, x_2, x_3, \ldots, x_N) = \Psi(x_2, x_1, x_3, \ldots, x_N)$$
$$= -\Psi(x_1, x_2, x_3, \ldots, x_N). \qquad (\text{III}.25)$$

EXERCISES

52. Show that the two-electron function

$$\Psi(x_1, x_2) = \Psi_a(x_1)\psi_b(x_2) - \psi_a(x_2)\psi_b(x_1)$$

is antisymmetric under the transposition $\mathbf{P}_{12}$. Is the product $\psi(x_2)\psi(x_1)$ antisymmetric?

53. **(a)** Which value does the function $\Psi(x_1, x_2)$ in Exercise 52 have when $x_1 = x_2$? **(b)** Notice that the function $\Psi(x_1, x_2)$ in Exercise 52 can be written as a determinant:

$$\Psi(x_1, x_2) = \begin{vmatrix} \psi_a(x_1) & \psi_b(x_1) \\ \psi_a(x_2) & \psi_b(x_2) \end{vmatrix}.$$

Notice carefully the labeling of the rows and columns. What

happens to a determinant if two rows or two columns change place? What is the value of a determinant with two identical columns or two identical rows? What is the value of a determinant obtained from the one above by letting rows and columns change places?

Many-Electron Hamiltonian

A neutral atom with nuclear charge Z has $N = Z$ electrons. The basic many-electron Hamiltonian for such a system is in atomic units with the Hartree as energy unit,

$$\mathbf{H} = -\frac{1}{2} \sum_{i=1}^{N} \Delta_i - Z \sum_{i=1}^{N} \frac{1}{r_i} + \frac{1}{2} \sum_{i \neq j=1}^{N} \frac{1}{r_{ij}}. \quad \text{(III.26)}$$

Here Δ_i is the Laplacian working on functions of $\mathbf{r}_i$, r_i is the distance between electron i and the nucleus, and r_{ij} is the distance between electrons i and j.

EXERCISES

54. Which physical quantities do the three sums in (26) represent?

55. Why are the terms with $i = j$ in the third sum of (26) excluded?

56. What is the reason for the factor $\frac{1}{2}$ in front of the third sum of (26)?

57. Why is there a minus sign in front of the second and a plus sign in front of the third sum of (26)?

The first two sums in (26) consist of one-electron operators and the last one of two-electron operators. This simple statement, which has far-reaching consequences, means that each term in the first two sums operates only on the coordinates of *one* electron, whereas each term in the third sum operates on *two* electrons. If it were not for the third sum, we would be able to write up the exact solutions of the problem directly. All approximations needed in order to approach accurate solutions of the many-electron problem have their root in the third sum of (26).

EXERCISES

58. Write up explicitly the Hamiltonian for the helium atom, which has two electrons.

59. Neglect the repulsion between the two electrons completely. How does the Hamiltonian look then?

60. Show that the product $\psi_{100}(\mathbf{r}_1)\psi_{100}(\mathbf{r}_2)$ [cf. (1)] forms an exact eigenfunction of the Hamiltonian in Exercise 59 when $\psi_{100}(\mathbf{r})$ is the ground-state solution of a hydrogen-atom-like problem. Why "hydrogen-atom-like," and not just "hydrogen atom"? Which total energy does such a "helium atom" have?

61. The product in Exercise 60 is not an acceptable wave function for the helium atom. Why?

Even though the many-electron Hamiltonian (26) does not contain any operator working on the spin coordinates, we need spin functions in order to handle the permutation symmetry correctly. For a summary of the basic aspects of spin coordinates and spin functions we refer to Section II.I. The simplest anti-symmetric wave functions—thus satisfying (24)—are determinants built up of spin orbitals. An *orbital* is a function of the vector $\mathbf{r}$, thus of the three "ordinary" spatial variables, and a *spin orbital* is a function of the combined ordinary space–spin space variable $x = (\mathbf{r}, \zeta)$. The simplest way of constructing spin orbitals is to multiply an orbital by one of the two possible spin functions, $\alpha(\zeta)$ or $\beta(\zeta)$. Given an orbital $\phi(\mathbf{r})$, we can thus construct the two spin orbitals $\psi_a(x) = \phi(\mathbf{r})\alpha(\zeta)$ and $\psi_b(x) = \phi(\mathbf{r})\beta(\zeta)$. From these we can construct a determinant

$$
\Psi(x_1, x_2) = \begin{vmatrix} \psi_a(x_1) & \psi_b(x_1) \\ \psi_a(x_2) & \psi_b(x_2) \end{vmatrix}
$$
$$
= \psi_a(x_1)\psi_b(x_2) - \psi_b(x_1)\psi_a(x_2)
$$
$$
= \phi(\mathbf{r}_1)\phi(\mathbf{r}_2)[\alpha(\zeta_1)\beta(\zeta_2) - \beta(\zeta_1)\alpha(\zeta_2)]. \qquad \text{(III.27)}
$$

This is a possible wave function for a two-electron system: the two electrons can occupy the same orbital provided that they have different spin functions.

EXERCISES

62. What happens with the spatial part $\phi(\mathbf{r}_1)\phi(\mathbf{r}_2)$ of the two-electron function (27) under permutation of the coordinates $\mathbf{r}_1$ and $\mathbf{r}_2$?

63. What happens with the spin part of (27) under permutation of the coordinates ζ_1 and ζ_2?

64. What happens with the total function $\Psi(x_1, x_2)$ in (27) under permutation of the coordinates x_1 and x_2?

We have seen that the product $\psi_{100}(\mathbf{r}_1)\psi_{100}(\mathbf{r}_2)$ constitutes an exact solution of the "helium atom" problem when the electronic repulsion is neglected. Despite that fact, such a product is not an acceptable wave function for the system. If we multiply it by the spin function $[\alpha(\zeta_1)\beta(\zeta_2) - \beta(\zeta_1)\alpha(\zeta_2)]$, we get, however, a two-electron function that is antisymmetric under permutations and therefore acceptable.

The situation illustrated here for a two-electron system is quite general. If the total Hamiltonian is a sum of one-electron operators, the Schrödinger equation is separable, i.e., the total wave function can be written as a product of one-electron functions—orbitals). To see that a little more explicitly, we introduce the notation [cf. (26)]

$$h_i = -\frac{1}{2}\Delta_i - \frac{Z}{r_i}; \qquad (III.28a)$$

$$\mathbf{H} = \sum_{i=1}^{N} h_i. \qquad (III.28b)$$

The product ansatz for the N-electron function (notice that we use a capital letter for the many-electron function and lowercase letters for the spin orbitals),

$$\Phi(x_1, x_2, x_3, \ldots, x_N) = \phi_a(x_1)\phi_b(x_2)\phi_c(x_3) \cdots \phi_n(x_N),$$

$$(III.29)$$

gives with (28*b*),

$$\mathbf{H}\Phi(x_1, x_2, x_3, \ldots, x_N)$$

$$= [h_1\phi_a(x_1)]\phi_b(x_2)\phi_c(x_3) \cdots \phi_n(x_N)$$

$$+ \phi_a(x_1)[h_2\phi_b(x_2)]\phi_c(x_3) \cdots \phi_n(x_N)$$

$$+ \phi_a(x_1)\phi_b(x_2)[h_3\phi_c(x_3)] \cdots \phi_n(x_N)$$

$$+ \cdots + \phi_b(x_2)\phi_c(x_3) \cdots [h_N\phi_n(x_N)]. \quad \text{(III.30)}$$

The one-electron Hamiltonian (28*a*) is, however, a hydrogen-like (why "like"?) operator and its eigenfunctions and eigenvalues are known. We can write

$$h_i\phi_k(x_i) = \epsilon_k\phi_k(x_i), \quad \text{(III.31)}$$

where the subscript k labels one of the eigenfunctions of the problem and ϵ_k is the corresponding energy eigenvalue.

EXERCISES

65. Write up the first nine eigenfunctions of the Hamiltonian (28*a*) and the corresponding energy eigenvalues. Are there any degeneracies?

66. How many electrons can occupy an *s*-state; a *p*-state; a *d*-state?

ATOMS

Combining (30) and (31), we get

$$\mathbf{H}\Phi(x_1, x_2, x_3, \ldots, x_N)$$
$$= [\epsilon_a + \epsilon_b + \epsilon_c + \cdots + \epsilon_n]\Phi(x_1, x_2, x_3, \ldots, x_N).$$

(III.32)

In other words, if the total Hamiltonian is a sum of one-electron operators, its eigenfunctions can be written as products of such spin orbitals, which are eigenfunctions of the one-electron operator, and its eigenvalues are sums of the corresponding one-electron eigenvalues. This is a very general statement and it does not tell us anything as to which spin orbitals should be used for a particular N-electron state.

EXERCISES

67. The lithium atom has three electrons. Neglect the interaction between the electrons and write up a ground-state three-electron function (product) and its corresponding energy. How many such products with the same—lowest possible—energy are there?

68. The carbon atom has six electrons. Which is its lowest possible energy if the interaction between the electrons is neglected? Is there more than one product function with that energy?

69. The neon atom has 10 electrons. Which is its lowest possible energy if the interaction between the electrons is neglected? Is there more than one product function with that energy?

44

The Antisymmetrizer

As we have seen earlier, a product function such as (29) cannot possibly be an acceptable wave function for an N-electron system, even though it may constitute an exact solution of a model problem with a Hamiltonian like (28*b*). To get wave functions that transform in the correct way under permutations of the electronic coordinates, we need determinants. In particular, we need a procedure to transform a product to a determinant. An operator called the *antisymmetrizer* does that. This operator is defined by

$$\mathbf{A} = \frac{1}{\sqrt{N!}} \sum_{\mathbf{P}}^{\text{all}} (-1)^p \, \mathbf{P}. \qquad \text{(III.33)}$$

This is thus a sum over all $N!$ permutations $\mathbf{P}$ of N objects, multiplied by a parity factor. The number in front of the sum is not necessary for the antisymmetry as such, but as we will see, it is practical to include this factor.

EXERCISE

70. In how many ways can two objects be permuted? Which are these permutations? How does the antisymmetrizer for $N = 2$ look? What happens when that operator works on the electronic coordinates of the functions (a) $\phi(x_1)\phi(x_2)$ and (b) $\psi(x_1)\phi(x_2)$?

Here we will need only three elementary properties of the antisymmetrizer. First we notice that a product of two permutations $\mathbf{P}$ and $\mathbf{Q}$ with parities p and q, respectively, is a permutation with parity $p + q$. We therefore have

$$\mathbf{QA} = \frac{1}{\sqrt{N!}} \sum_{\mathbf{P}}^{\text{all}} (-1)^p \, \mathbf{QP} = \frac{1}{\sqrt{N!}} \sum_{\mathbf{P}}^{\text{all}} (-1)^{p+2q} \mathbf{QP}$$

$$= (-1)^q \frac{1}{\sqrt{N!}} \sum_{\mathbf{P}}^{\text{all}} (-1)^{p+q} \, \mathbf{QP}$$

$$= (-1)^q \frac{1}{\sqrt{N!}} \sum_{\mathbf{r}}^{\text{all}} (-1)^r \, \mathbf{R} = (-1)^q \, \mathbf{A}. \qquad \text{(III.34)}$$

Here we have thus used the fact that the sum over all the $N!$ permutations $\mathbf{QP}$ with their parities $p + q$, when $\mathbf{Q}$ is a fixed permutation and $\mathbf{P}$ runs through all permutations can be written as a sum over the permutations $\mathbf{R} = \mathbf{QP}$ with parities $r = p + q$. We also notice the form of (34), which can be thought of as an "eigenoperator relation": the antisymmetrizer is an eigenoperator of any permutation with an eigenvalue that is $+1$ for even and -1 for odd permutations. We can apply (34) to an arbitrary function of N variables $f(x_1, x_2, x_3, \ldots, x_N)$,

$$\mathbf{A}f(x_1, x_2, x_3, \ldots, x_N) = F(x_1, x_2, x_3, \ldots, x_N).$$

(III.35)

Combining (34) and (35), we see that the functions F satisfy the relation

$$\mathbf{Q}F(x_1, x_2, x_3, \ldots, x_N) = (-1)^q F(x_1, x_2, x_3, \ldots, x_N),$$

(III.36)

that is, $F(x_1, x_2, x_3, \ldots, x_N)$ is antisymmetric under permutations of the variables x_i. We should add that the function $f(x_1, x_2, x_2, \ldots, x_N)$ on which the antisymmetrizer operates can be such that the result vanishes. But if the result is different from zero, it is an antisymmetric function, as shown by (36).

Next we are going to show that apart from a constant factor the square of the antisymmetrizer is equal to itself. Using the definition (33) and the same reasoning as in (34), we have

$$\mathbf{A}^2 = \frac{1}{N!} \sum_{P,Q}^{\text{all}} (-1)^{p+q} \mathbf{PQ} = \frac{1}{N!} \sum_{P}^{\text{all}} \sum_{R}^{\text{all}} (-1)^r \mathbf{R}$$

$$= \frac{1}{\sqrt{N!}} \sum_{P}^{\text{all}} \mathbf{A} = \sqrt{N!} \, \mathbf{A}. \qquad \text{(III.37)}$$

Thus to operate with the antisymmetrizer twice on a function is equivalent to operating once and multiplying by $\sqrt{N!}$. This rule will be useful when we calculate matrix elements with respect to determinants.

71. Verify (37) for the case $N = 2$.

Finally, we show that the antisymmetrizer is a Hermitian operator. This may at first seem surprising, since the permutations themselves are in general *not* Hermitian. That can be seen by the following counterexample $[dx_1, dx_2, dx_3 = (dx)]$:

$$\mathbf{P}_{123}\Psi(x_1, x_2, x_3) = \Psi(x_2, x_3, x_1); \quad (\text{III.}38a)$$

$$\mathbf{P}_{132}\Psi(x_1, x_2, x_3) = \Psi(x_3, x_1, x_2); \quad (\text{III.}38b)$$

$$\int (dx)\, \Phi^*(x_1, x_2, x_3)\, \mathbf{P}_{123}\Psi(x_1, x_2, x_3)$$

$$= \int (dx)\, \Phi^*(x_1, x_2, x_3)\Psi(x_2, x_3, x_1); \quad (\text{III.}38c)$$

$$\int (dx)\, \Phi^*(x_1, x_2, x_3)\mathbf{P}_{123}^+\Psi(x_1, x_2, x_3)$$

$$= \int (dx)\, [\mathbf{P}_{123}\Phi(x_1, x_2, x_3)]^*\Psi(x_1, x_2, x_3)$$

$$= \int (dx)\, \Phi^*(x_2, x_3, x_1)\Psi(x_1, x_2, x_3)$$

$$= \int (dx')\, \Phi^*(x_1', x_2', x_3')\Psi(x_3', x_1', x_2')$$

$$= \int (dx')\, \Psi^*(x_1', x_2', x_3')\, \mathbf{P}_{132}\Psi(x_1', x_2', x_3') \quad (\text{III.}38d)$$

72. Which is the adjoint operator to $\mathbf{P}_{123}$? Which operator is adjoint to $\mathbf{P}_{132}$?

73. Is $\mathbf{P}_{12}$ Hermitian?

74. Which are the inverses of $\mathbf{P}_{12}$, $\mathbf{P}_{123}$, and $\mathbf{P}_{132}$?

The antisymmetrizer contains all the $N!$ permutations of N objects. This means that for every permutation $\mathbf{P}$, its inverse, $\mathbf{P}^{-1}$, is also included. The antisymmetrizer is therefore Hermitian:

$$\mathbf{A}^{+} = \mathbf{A}. \qquad (III.39)$$

The properties (37) and (39) are useful in connection with the turnover rule (see Section II.H). If we want to calculate the matrix element of an operator $\mathbf{F}$ that commutes with all the permutations [(28*b*) provides an example], with respect to two antisymmetric functions, we can proceed as follows:

$$\int (dx) \, [\mathbf{A} \, \Phi(x_1, x_2, x_3, \ldots, x_N)]^{*}$$

$$\cdot \, \mathbf{FA}\Psi(x_1, x_2, x_3, \ldots, x_N)$$

$$= \int (dx) \, \Phi^{*}(x_1, x_2, x_3, \ldots, x_N)$$

$$\cdot \, \mathbf{A}^{+}\mathbf{FA}\Psi(x_1, x_2, x_3, \ldots, x_N)$$

$$= \int (dx)\, \Phi^*(x_1, x_2, x_3, \ldots, x_N)$$

$$\cdot \mathbf{AFA}\Psi(x_1, x_2, x_3, \ldots, x_N)$$

$$= \int (dx)\, \Phi^*(x_1, x_2, x_3, \ldots, x_N)$$

$$\cdot \mathbf{FA}^2\Psi(x_1, x_2, x_3, \ldots, x_N)$$

$$= \sqrt{N!} \int (dx)\, \Phi^*(x_1, x_2, x_3, \ldots, x_N)$$

$$\cdot \mathbf{FA}\Psi(x_1, x_2, x_3, \ldots, x_N). \tag{III.40}$$

EXERCISES

75. How many terms do the first and last lines of (40), respectively, contain? What are those numbers for $N = 3, 5, 10$?

76. In the case $N = 2$, let $\Phi(x_1, x_2) = \Psi(x_1, x_2) = f(x_1)g(x_2)$. Calculate explicitly the first and last lines of (40). How is it possible that the two expressions can be equal?

Slater's Rules

Nearly all methods sooner or later use spin orbitals, one-electron functions of the combined "ordinary space"–spin space variable $x = (\mathbf{r}, \zeta)$: $f(x)$. Since the total wave function for a many-electron system must be antisymmetric, the natural basis functions for many-electron systems are antisymmetrized products of spin orbitals—in other words, determinants. The many-electron Hamiltonian contains sums of one- and two-electron operators. Matrix elements of such operators with respect to deter-

ATOMS

minants can be reduced to combinations of integrals over spin orbitals. At the end of the 1920s, Slater derived rules for doing that in the case when the spin orbitals are orthonormal. Löwdin extended these rules to the general case in 1955. Here we work only with determinants built up of orthonormal spin orbitals, thus satisfying

$$\int dx\, f_k^*(x) f_l(x) = \delta_{kl}. \tag{III.41}$$

EXERCISES

77. A reasonable choice of spin orbitals for the helium atom is obtained by combining the $1s$ and $2s$ hydrogen-like orbitals (look them up to get explicit expressions), multiplied by the two spin functions $\alpha(\zeta)$ and $\beta(\zeta)$. Check that the $1s$ orbital is orthogonal to the $2s$ orbital (notice how these terms are used), and recall also what is meant by saying that the two spin functions are orthogonal to each other.

78. Are the following *spin orbitals* orthogonal, and if yes, why? (a) $1s\alpha$ and $1s\beta$; (b) $1s\alpha$ and $2s\alpha$; (c) $2s\beta$ and $1s\alpha$.

Slater's rules can be derived very simply by means of the relations (37) and (39) for the antisymmetrizer. We start with the overlap integral between two antisymmetrized products,

$$\Psi_K(x_1, x_2, x_3, \ldots, x_N)$$

$$= \mathbf{A}\,[f_{k_1}(x_1) f_{k_2}(x_2) f_{k_3}(x_3) \cdots f_{k_N}(x_N)]$$

$$= \mathbf{A}\Phi_K(x_1, x_2, x_3, \ldots, x_N); \tag{III.42a}$$

50

$$\Psi_L(x_1, x_2, x_3, \ldots, x_N)$$

$$= \mathbf{A}\,[f_{1_1}(x_1) f_{1_2}(f_{1_3}(x_3) \cdots f_{1_N}(x_N)]$$

$$= \mathbf{A}\Phi_L(x_1, x_2, x_3, \ldots, x_N). \tag{III.42b}$$

Notice that we use the letters Φ for the products and Ψ for the antisymmetrized products (i.e., the determinants). In both cases the capital letters K and L denote the entire set of lowercase subscripts k_i or l_j.

We will use the same technique as in (40) but express it in a more concise form by means of the Dirac notation

$$\langle \Psi_K | \Psi_L \rangle = \langle \mathbf{A}\Phi_K | \mathbf{A}\Phi_L \rangle = \langle \Phi_K | \mathbf{A}^+ \mathbf{A}\Phi_L \rangle$$

$$= \langle \Phi_K | \mathbf{A}^2 \Phi_L \rangle = \sqrt{N!}\,\langle \Phi_K | \mathbf{A}\Phi_L \rangle$$

$$= \sum_{\mathbf{P}}^{\text{all}} (-1)^p \langle \Phi_K | \mathbf{P}\Phi_L \rangle. \tag{III.43}$$

EXERCISES

79. How many terms are there in the first three expressions in (43)? How many in the final expression?

80. What is the reason for the reduction of the number of terms in (43)?

For the first term in the sum over permutations in (43), $\mathbf{P} = \mathbf{1}$, we get

$$\langle \Phi_K | \Phi_L \rangle_{\mathbf{P}=\mathbf{1}} = \delta_{k_1 l_1} \delta_{k_2 l_2} \delta_{k_3 l_3} \cdots \delta_{k_N l_N}. \tag{III.44}$$

In other words, (44) vanishes unless each spin orbital in the first product matches its corresponding spin orbital in the second product.

---

81. Carry through the intermediate steps that lead to the final result in (44). Why do the k and l subscripts pair off in that particular way?

The remaining terms in (43) with permutations other than the identity will all give zero. If the "l-set" is identical to the "k-set," a permutation will unavoidably introduce a "mismatch," which will yield one or more factors equal to zero. (Why can't the sets contain two or more identical spin orbitals such that certain permutations would have no effect?) If the l-set is different from the k-set of spin orbitals, there will be even more possibilities to get zero.

Thus we have the first Slater rule: two determinants built up of orthonormal spin orbitals are orthogonal (as N-electron functions) unless they are identical (apart from a possible change of sign). When they are identical, the previous discussion also shows that the function

$$D_K(x_1, x_2, x_3, \ldots, X_N)$$

$$= \mathbf{A}\,[f_{k_1}(x_1)\, f_{k_2}(x_2)\, f_{k_3}(x_3) \cdots f_{k_N}(x_N)]$$

$$= \frac{1}{\sqrt{N!}} \begin{vmatrix} f_{k_1}(x_1) & f_{k_2}(x_1) & f_{k_3}(x_1) & \cdots & f_{k_N}(x_1) \\ f_{k_1}(x_2) & f_{k_2}(x_2) & f_{k_3}(x_2) & \cdots & f_{k_N}(x_2) \\ \cdots & \cdots & \cdots & \cdots \cdots \\ f_{k_1}(x_N) & f_{k_2}(x_N) & f_{k_3}(x_N) & \cdots & f_{k_N}(x_N) \end{vmatrix}$$

$$= \frac{1}{\sqrt{N!}}\, \det\,\{ f_{k_i}(x_j) \} \tag{III.45}$$

is normalized.

82. Write up the antisymmetrizer for $N = 2$ explicitly. Apply it to the function $f(\mathbf{r}_1)\alpha(\zeta_1)f(\mathbf{r}_2)\beta(\zeta_2)$. Normalize that function assuming that the orbital $f(\mathbf{r})$ itself is normalized.

83. Apply the antisymmetrizer to the function $f(\mathbf{r}_1)\alpha(\zeta_1)f(\mathbf{r}_2)\alpha(\zeta_2)$. Explain the result.

84. Apply the antisymmetrizer to the function $f(\mathbf{r}_1)\alpha(\zeta_1)g(\mathbf{r}_2)\beta(\zeta_2)$. Calculate the overlap integral between the result and the determinant obtained in Exercise 82.

We then illustrate the derivation of Slater's rules with another case, in which we want to simplify the many-electron integral $\langle D|\Omega|D_k^\mu\rangle$, where D is a determinant like (45) with the spin orbitals f_{k_j}, $j = 1, 2, 3, \ldots, N$, and D_k^μ is obtained from it by replacing the spin orbital f_k by another one, f_μ, which is orthonormal to all the spin orbitals in D. We use F_k^μ and D_k^μ for the corresponding products so that $D = \mathbf{A}F$ and $D_k^\mu = \mathbf{A}F_k^\mu$. The operator Ω is a symmetric sum of one-electron operators:

$$\Omega = \sum_{i=1}^{N} \Omega_i. \tag{III.46}$$

The notation in (46) means that Ω_i works only on functions of the variable x_i

85. Let Ω_i be the momentum operator for electron i. Write out $\Omega_1 = p_1$ and $\Omega_3 = p_3$ explicitly.

86. How, then, can the sum Ω be interpreted?

87. Calculate **(a)** $p_1 f(x_1)f(x_2)$; **(b)** $p_1 f(x_1)g(x_2)$, **(c)** $p_1 f(x_3)f(x_7)$, and **(d)** $[p_1 + p_2]f(x_1)f(x_2)$.

The fact that Ω is (46) is a symmetric sum implies that for any permutation **P** of the electronic coordinates x_i,

$$[\mathbf{P}, \Omega] = 0. \qquad (\text{III.47})$$

Consequently, the antisymmetrizer (33) also commutes with (46):

$$[\mathbf{A}, \Omega] = 0. \qquad (\text{III.48})$$

Proceeding as in (43), we then get

$$\langle D|\Omega|D_k^\mu\rangle = \langle \mathbf{A}F|\Omega|\mathbf{A}\,F_k^\mu\rangle = \langle F|\,\mathbf{A}^+\Omega|\mathbf{A}\,F_k^\mu\rangle$$

$$= \langle f|\,\mathbf{A}\Omega|\mathbf{A}\,F_k^\mu\rangle$$

$$= \langle f|\Omega|\mathbf{A}^2 F_k^\mu\rangle = \sqrt{N!}\,\langle F|\Omega|\mathbf{A}F_k^\mu\rangle$$

$$= \sum_{\mathbf{P}}^{\text{all}} (-1)^p \langle F|\Omega|\mathbf{P}F_k^\mu\rangle. \qquad (\text{III.49})$$

88. Motivate each of the various steps in (49)

89. How many terms do the various intermediate sums in (49) contain when $N = 2$?

When we consider the first term Ω_1 in the operator sum in (46), a typical term in the last sum of (49) is

$$(-1)^P \langle F|\Omega_1|\mathbf{P}F_k^\mu\rangle$$

$$= (-1)^P \langle 1|\Omega_1|1'\rangle\langle 2|2'\rangle \cdots \langle N|N'\rangle. \quad \text{(III.50)}$$

Here we thus used primed variables to indicate the result of the permutation **P**. In general, this product of overlap integrals will contain at least one that vanishes. (Why?) The only chance to avoid that is to let f_k and f_μ match each other in an integral that also contains the operator Ω_1:

$$\langle k|\Omega_1|\mu\rangle = \int dx\, f_k^*(x_1)\Omega_1 f_\mu(x_1). \quad \text{(III.51)}$$

This argument is also valid for the other terms in the operator sum (46), and therefore the final result is

$$\langle D|\Omega|D_k^\mu\rangle = \langle k|\Omega_1|\mu\rangle. \quad \text{(III.52)}$$

It is very important to realize, first, what the notations in (52) stand for. In particular, we notice that the subscript "1" on the operator on the right-hand side of (52) could be called anything, since it is a so-called dummy variable. The fact that the left-hand side of (52) is a sum of N-electron integrals and the right-hand side is a single one-electron integral is indeed impressive.

90. Carry through a direct calculation of the integral (49) in the case $N = 2$. Set $f_1(x) = f(x)$, $f_2(x) = g(x)$, and $f_\mu(x) = h(x)$. Thus simplify the two-electron integral $\langle D|\Omega|D_2^\mu\rangle$.

91. Calculate $\langle D|\Omega|D_1^\mu\rangle$.

92. Use these two examples to illustrate the reasoning in the derivation of the general case. Do the results agree?

All Slater's rules for simplifying expectation values of many-electron operators with respect to many-electron functions can be derived in a similar way. We now summarize these rules.

I. Overlap integrals

$$\langle D_K|D_L\rangle = \pm\, \delta_{k_1 l_1}\delta_{k_2 l_2}\delta_{k_3 l_3} \cdots \delta_{k_N l_N}. \qquad \text{(III.53)}$$

II. Integrals over sums of one-electron operators
   a. Diagonal elements:

$$\langle D|\sum_{i=1}^{N} \Omega_i\,|D\rangle = \sum_{k=1}^{N} \langle k|\,\Omega_1\,|k\rangle. \qquad \text{(III.54}a)$$

   b. The case $f_k \rightarrow f_\mu$:

$$\langle D|\sum_{i=1}^{N} \Omega_i\,|D_k^\mu\rangle = \langle k|\Omega_1\,|\mu\rangle. \qquad \text{(III.54}b)$$

   c. Zero in all other cases.

III. Integrals over sums of two-electron operators
    a. Diagonal elements:

$$\langle D| \sum_{i<j}^{N} \Omega_{ij} |D\rangle = \sum_{k<1}^{N} \langle kl | \Omega_{12} (\mathbf{1} - \mathbf{P}_{12}) |kl\rangle .$$

(III.55*a*)

    b. The case $f_l \rightarrow f_\mu$:

$$\langle D| \sum_{i<j}^{N} |D_l^\mu\rangle = \sum_{k=1}^{N} \langle kl | \Omega_{12} (\mathbf{1} - \mathbf{P}_{12}) |k\mu\rangle .$$

(III.55*b*)

    c. The case $f_k \rightarrow f_\mu, f_l \rightarrow f_v$:

$$\langle D| \sum_{i<j}^{N} \Omega_{ij} | D_{kl}^{\mu v}\rangle = \langle kl | \Omega_{12} (\mathbf{1} - \mathbf{P}_{12}) |\mu v\rangle .$$

(III.55*c*)

    d. Zero in all other cases.

In (55), $\mathbf{P}_{12}$ is a transposition that interchanges x_1 and x_2. For example, the final integral in (55*c*) is

$$\langle kl | \omega_{12} (\mathbf{1}-\mathbf{P}_{12}) |\mu v\rangle$$

$$= \int dx_1 \, dx_2 \, f_k^*(x_1) f_l^*(x_2) \Omega_{12}(\mathbf{1} - \mathbf{P}_{12}) f_\mu(x_1) f_v(x_2)$$

$$= \int dx_1 \, dx_2 \, f_k^*(x_1) f_l^*(x_2) \Omega_{12} f_\mu(x_1) f_v(x_2)$$

$$- \int dx_1 \, dx_2 \, f_k^*(x_1) f_l^*(x_2) \Omega_{12} f_v(x_1) f_\mu(x_2)$$

$$= \langle kl | \Omega_{12} |\mu v\rangle - \langle kl | \Omega_{12} |v\mu\rangle .$$

(III.56)

The most common two-electron operator is the electron repulsion $\Omega_{12} = 1/r_{12}$. In that case one often uses the Mulliken notation for these two-electron integrals:

$$\langle kl \mid \frac{1}{r_{12}} \mid \mu v \rangle = \int dx_1 \, dx_2 \, \frac{f_k^*(x_1) f_l^*(x_2) f_\mu(x_1) f_v(x_2)}{r_{12}}$$

$$= (k\mu \mid lv). \tag{III.57}$$

The Mulliken notation emphasizes the point that such an integral can be interpreted as the electrostatic interaction between the two charge distributions $f_k^*(x_1) f_\mu(x_1)$ and $f_l^*(x_2) f_v(x_2)$.

EXERCISES

93. Show that the double sum over the two-electron operators can be written either as

$$\sum_{i<j}^{N} \frac{1}{r_{ij}} \quad \text{or} \quad \frac{1}{2} \sum_{i,j}^{N}{}' \frac{1}{r_{ij}},$$

where the prime on the second sum indicates that the terms with $i = j$ should be omitted.

94. Show that the double sum over spin orbitals in (55a) can also be written in two ways:

$$\sum_{k<l}^{N} \langle kl \mid \Omega_{12} (1 - P_{12}) \mid kl \rangle \quad \text{or} \quad \tfrac{1}{2} \sum_{k,l}^{N} \langle kl \mid \Omega_{12} (1 - P_{12}) \mid kl \rangle,$$

and notice that the reason is not quite the same as in Exercise 93.

95. Show that the following symmetry rules hold for two-electron integrals: **(a)** $(kl \mid mn) = (mn \mid kl)$ and **(b)** $(kl \mid mn) = (lk \mid nm)^*$.

Densities and Form Factors for Many-Electron Atoms

We are now in a position to extend the discussion of densities and form factors from the one-electron case treated in section III.A to many-electron atoms. The density associated with a wave function can be defined in several different but equivalent ways. Here we define the density as the expectation value of the following sum of Dirac delta functions interpreted as one-electron operators.

$$\rho_{op} = \sum_{i=1}^{N} \delta(\mathbf{r} - \mathbf{r}_i). \qquad \text{(III.58)}$$

EXERCISES

96. Apply (58) to the one-electron case. Does the result agree with (17)?

97. Is the expression (58) intuitively reasonable?

An N-electron system characterized by the wave function $\Psi(x_1, x_2, x_3, \ldots, x_N)$ then has the density

$$\rho(\mathbf{r}) = \int dx_1\, dx_2\, dx_3 \cdots dx_N \Psi^*(x_1, x_2, x_3, \ldots, X_N)$$

$$\cdot\, \rho_{op}\Psi(x_1, x_2, x_3, \ldots, x_N). \qquad \text{(III.59)}$$

If the total wave function is single Slater determinant such as (45), we can use Slater's rule (54a) to write the density as

$$\rho(\mathbf{r}) = \sum_{k=1}^{N} \int dx_1\, f_k^*(x_1)\delta(\mathbf{r} - \mathbf{r}_1)f_k(x_1). \qquad \text{(III.60)}$$

59

The spin orbitals $f_k(x)$ are often—but by no means always—products of an orbital and a spin function:

$$f_1(x) = \phi_1(\mathbf{r})\alpha(\zeta); \qquad f_2(x) = \phi_1(\mathbf{r})\beta(\zeta);$$

$$f_3(x) = \phi_2(\mathbf{r})\alpha(\zeta); \qquad f_4(x) = \phi_2(\mathbf{r})\beta(\zeta);$$

$$\cdots \qquad\qquad \cdots$$

$$f_{2\nu-1}(x) = \phi_\nu(\mathbf{r})\alpha(\zeta); \qquad f_{2\nu}(x) = \phi_\nu(\mathbf{r})\beta(\zeta);$$

$$\cdots \qquad\qquad \cdots \qquad\qquad \text{(III.61)}$$

The spin integration in (60) gives 1 (Why?) and the density for a system with an even number of electrons is

$$\rho(\mathbf{r}) = 2 \sum_{\nu=1}^{N/2} |\phi_\nu(\mathbf{r})|^2. \qquad \text{(III.62)}$$

This expression is easily interpreted: each doubly filled orbital contributes twice its absolute value squared to the density. Such a direct interpretation is possible only when the total wave function is a single determinant and when the spin orbitals in that determinant are of the form (61).

98. Calculate the density of a two-electron system described by a single determinant built up of the spin orbitals $f(\mathbf{r})\alpha(\zeta)$ and $g(\mathbf{r})\beta(\zeta)$, **(a)** directly and **(b)** using (60). Do the results agree?

99. Use (60) to calculate the density of a three-electron atom when the total wave function is a single determinant with the spin orbitals $f(\mathbf{r})\alpha(\zeta)$, $f(\mathbf{r})\beta(\zeta)$, and $g(\mathbf{r})\alpha(\zeta)$.

For an atom the generic subscript ν that we have used in (61) and (62) is a compound subscript consisting of a radial quantum number n and two angular quantum numbers l and m.

EXERCISES

100. Use the notation (nlm) to list those atomic orbitals which are occupied—once or twice—in the following atoms: **(a)** He, **(b)** Li, **(c)** C, **(d)** O, and **(e)** Ne.

101. Which are the corresponding densities?

When there are enough electrons to fill closed shells we can simplify the expression for the density further. The term *closed shell* here means that for a particular value of l, all the orbitals with $m = -l, -l + 1, -l + 2, \ldots, l - 1, l$, are filled. In such a case we can use the following result:

$$\sum_{m=-l}^{l} Y_{lm}(\vartheta, \varphi) Y_{lm}^{*}(\vartheta, \varphi) = \frac{2l + 1}{4\pi}. \qquad (\text{III.63})$$

EXERCISES

102. Look up the explicit expressions for the three spherical harmonics with $l = 1$, and verify (63) in that case.

103. Calculate the density for the beryllium atom using (63) when the occupied orbitals are written as

$$R_{10}(r)Y_{00}(\vartheta, \varphi) \quad \text{and} \quad R_{20}(r)Y_{00}(\vartheta, \varphi).$$

104. Use (63) to calculate the density for the neon atom when the occupied orbitals are written as

$$R_{10}(r)Y_{00}(\vartheta, \varphi), \ R_{20}(r)Y_{00}(\vartheta, \varphi) \quad \text{and} \quad R_{21}(r)Y_{11}(\vartheta, \varphi),$$

$$R_{21}(r)Y_{10}(\vartheta, \varphi), \ R_{21}(r)Y_{1(-1)}(\vartheta, \varphi).$$

For a many-electron atom with closed shells we can thus write the density (62) as

$$\rho(\mathbf{r}) = 2 \sum_{n=1}^{occ} \sum_{l=0}^{occ} \sum_{m=l}^{l} R_{nl}^2(r)|Y_{lm}(\vartheta, \varphi)|^2$$

$$= \frac{1}{2\pi} \sum_{n=1}^{occ} \sum_{l=0}^{occ} (2l + 1)R_{nl}^2(r). \tag{III.64}$$

EXERCISES

105. Which are the upper limits for the summations over n and l in (64) for the following atoms or ions? **(a)** He; **(b)** Ne; **(c)** Ar; **(d)** F; **(e)** K^+; **(f)** Br^-; **(g)** Cs^+.

106. What can be said about the angular dependence of the density in an atom with only closed shells?

Most atoms and ions have both closed and open shells, and then (63) must be combined with the general expression.

---

107. Simplify as much as possible, using notation similar to that above, the density of the ground states of the following atoms: **(a)** Li; **(b)** Be; **(c)** B; **(d)** C; **(e)** N; **(f)** O; **(g)** F.

108. Are any of these densities spherically symmetric?

The expression (62) is based on the assumption that the total wave function for the system in the state under investigation can be approximated by a single determinant built up of doubly or singly filled orbitals. When the atom contains one or more open shells, more than one determinant may be a possible candidate. Then it is time to question the starting assumption about the total wave function. Symmetry—primarily angular momentum symmerty—is then the first tool that one needs. But one must also make a general analysis of the overall description of the system. To do that falls outside the scope of this book.

Momentum Distribution

When the total wave function is approximated by a single determinant, each occupied orbital in position space has a counterpart in momentum space [cf. (5)]. For an atomic system these orbitals can be written [cf. (10)]

$$\phi(\mathbf{r}) = R_{nl}(r)Y_{lm}(\vartheta, \varphi);$$

$$\underline{\phi}(\mathbf{p}) = \underline{R}_{nl}(p)\underline{Y}_{lm}(\vartheta_p, \varphi_p). \qquad \text{(III.65)}$$

109. Use (10) to write up explicit expressions for $\underline{R}_{10}(p)$ and $\underline{R}_{21}(p)$ without carrying out the final integrations.

The momentum distribution *corresponding* to the density (62) is thus

$$\underline{\rho}(\mathbf{p}) = 2 \sum_{\nu=1}^{N/2} |\underline{\phi}_\nu(\mathbf{p})|^2. \qquad \text{(III.66)}$$

It is very important to notice that this momentum distribution *cannot* be calculated from the density $\rho(\mathbf{r})$. We can either calculate the momentum-space orbitals $\underline{\phi}_\nu(\mathbf{p})$ directly from their position-space counterparts, or use the following generalization of (19):

$$\underline{\rho}(\mathbf{p}) = \frac{2}{8\pi^3} \sum_{\nu=1}^{N/2} \int dv \, dv' \, \phi_\nu(\mathbf{r}) \phi_\nu^*(\mathbf{r}') e^{i\mathbf{p}\cdot(\mathbf{r}'-\mathbf{r})}. \qquad \text{(III.67)}$$

110. What is the difference between the density $\rho(\mathbf{r})$ and the expression used in (67)?

111. Is there any connection between (66) and (67)?

As seen from (65), two corresponding atomic orbitals $\phi_v(\mathbf{r})$ and $\underline{\phi}_v(\mathbf{p})$ have the same angular symmetry. The arguments about closed shells therefore hold also for the momentum distribution.

112. Which of the following atoms have a spherically symmetric momentum distribution in their ground states? **(a)** He; **(b)** Be; **(c)** C; **(d)** Ne.

113. Express the momentum distribution for the neon atom in terms of its radial momentum functions.

Form Factor

The general expression (20) can be used directly for atomic densities. Together with the expression for a plane wave in spherical harmonics (8), we can obtain a number of explicit results rather easily.

114. What can be said about the form factor for an atom with a spherically symmetric density?

115. Express the form factors for He and Ne in their ground states in terms of their radial functions.

116. Simplify the form factor for the boron atom as much as possible.

The atomic form factors are essential ingredients in the interpretation of diffraction experiments. What has been said here about atomic form factors applies as well [cf. (22)] to the reciprocal form factors.

EXERCISES

117. What can be said about the reciprocal form factor for an atom with a spherically symmetric momentum distribution?

118. Express the reciprocal form factor for He and Ne in their ground states in terms of their radial momentum functions.

It is important to notice that many of the results derived in this chapter are by no means restricted to atoms. This concerns, in particular, the sections about antisymmetry and Slater's rules.

REFERENCES

Atkins, P. W., *Molecular Quantum Mechanics*, Oxford University Press, Oxford, 1970; second edition, 1983.

Landau, L. D. and E. M. Lifshitz, *Quantum Mechanics*, Pergamon Press, Oxford, 1958.

Löwdin, P.-O., *Adv. Phys.* **5,** 1 (1956).

Messiah, A., *Mécanique quantique*, Dunod, Paris, 1959; English translation *Quantum Mechanics*, Wiley, New York, 1961.

IV

SMALL MOLECULES

General Aspects

A molecule is a collection of several nuclei and electrons. Even though the term *molecule* could in principle be used for any number of nuclei, it is important—as we will see in later chapters—to distinguish between *small molecules* with a finite number of nuclei on one hand and *extended systems* on the other, where in a certain sense one works with an infinite number of nuclei. In the present chapter we discuss some fundamental aspects of the theory of small molecules.

The basic difference between atoms and molecules is thus the presence of more than one nucleus in a molecule. That has far-reaching consequences for the molecular symmetry properties. But to begin with, we just notice that in a molecule there is in general no natural way to refer the motions of all the particles to *one* center. The fact that there is more than one nucleus also implies that the relative motions of the various nuclei are important. The electronic structure of the molecule is obviously essential, and in many cases the interaction between electronic and nuclear motion must also be taken into account. Fortunately, the very different masses of nuclei and electrons make it possible to separate approximately the treatment of nuclei and electrons and provide the background for the primary approximation in molecular science—the *Born–Oppenheimer* (BO) *approximation*.

EXERCISES

1. What are the ratios between the masses of **(a)** the proton and the electron; **(b)** the deuteron and the electron; and **(c)** the normal carbon nucleus and the electron?

2. How many nuclei and electrons do the following molecules contain? **(a)** H_2; **(b)** O_2; **(c)** C_2H_4; **(d)** benzene; **(e)** ammonia; **(f)** water.

The BO-approximation means essentially that we study the motion of the electrons in the presence of a frozen nuclear skeleton. Another way of expressing that is to say that we neglect the kinetic energy of the nuclei. The appearance of the nuclear masses in the denominators of the terms representing the nuclear kinetic energy provides an intuitive reason for such an approximation. The "physical" motivation for the BO-approximation is thus the fact that the electrons move much faster than the nuclei. It therefore makes some sense to study the electronic structure in the presence of temporarily fixed nuclei. This does not mean, however, that the motion of the nuclei is neglected completely, only that the electronic structure is assumed to follow the nuclear motion adiabatically. We will not go into any derivation or justification of the BO-approximation here, but the discussion throughout the chapter is based on that approximation.

General BO Hamiltonian and Many-Electron Wave Functions for a Small Molecule

The molecular counterpart to (III.26) is

$$\mathbf{H} = \frac{1}{2} \sum_{g,h}^{M} {}' \frac{Z_g Z_h}{R_{gh}} - \frac{1}{2} \sum_{i=1}^{N} \Delta_i - \sum_{g}^{M} \sum_{i}^{N} \frac{Z_g}{r_{gi}} + \frac{1}{2} \sum_{i,j}^{N} {}' \frac{1}{r_{ij}}. \quad \text{(IV.1)}$$

This is the Hamiltonian in the BO-approximation for a system of M nuclei with charges Z_g and N electrons, in atomic units with the hartree as the energy unit. The nuclei are labeled with the subscripts g and h, and the electrons with the subscripts i and j.

EXERCISES

3. Which relation holds between N, M, and the nuclear charges Z_g for a neutral molecule?

4. Write up explicit expressions for the distances R_{gh}, r_{ig}, and r_{ij} in terms of the coordinates of the individual nuclei and electrons.

5. The hartree, which is one of the so-called atomic units for energy, is obtained when the electronic mass and the electronic charge (its absolute value) are set equal to 1, and Planck's constant is set equal to 2π. The ionization potential of the hydrogen atom in its ground state is 0.5 hartree. How much is a hartree in the following energy units? (a) Electron volts; (b) cm^{-1}; (c) kcal/mol.

It is particularly important to notice the first sum in (1), which represents the energy due to the repulsion between the positively charged nuclei. In the BO-approximation the distances $R_{gh} = |\mathbf{R}_g - \mathbf{R}_h|$ between the nuclei are constants and the entire nuclear repulsion energy is thus a constant. As such, it presents no problem in the ensuing studies of the electronic structure. It means only that to have an electrically neutral system, we must include that constant term in the total energy. As we will see in later chapters, this becomes crucial in the treatment of extended systems.

6. The nuclear kinetic energy is thus neglected in (1). How does that term look explicitly?

7. Write up the complete (thus including the nuclear kinetic energy) many-particle Hamiltonian explicitly for **(a)** H_2; **(b)** O_2; **(c)** the water molecule; and **(d)** the methane molecule.

Even with a frozen nuclear skeleton, the many-electron Schrödinger equation

$$\mathbf{H}\Psi = E\Psi \qquad \text{(IV.2)}$$

for a small molecule presents a formidable problem, which can only be treated approximately. What is very important to notice, however, is that those approximations that have been developed so far provide useful and efficient tools for treating a very broad range of realistic chemical and physical problems. The methods for constructing approximate solutions of (2) are continuously being improved from both the point of view of formal mathematics and of their implementation as computer programs.

The many-electron molecular wave function in (2) depends on all the electronic coordinates $x_i = (\mathbf{r}_i, \zeta_i)$ as "ordinary" variables *and* on the nuclear coordinates $\mathbf{R}_g$ as parameters:

$$\Psi = \Psi(\mathbf{R}_1, \mathbf{R}_2, \ldots \mathbf{R}_g, \ldots \mathbf{R}_M;$$
$$x_1, x_2, \ldots, x_i, \ldots, x_N). \qquad \text{(IV.3)}$$

This means that for each set of nuclear coordinates $\{\mathbf{R}_g\}$ we have one wave function in the state under consideration as a function of the electronic coordinates $\{x_i\}$. In order not to burden the notation, we will, however, suppress the nuclear coordinates in most expressions from now on.

IV.A. THE HYDROGEN MOLECULE

Generalities

The hydrogen molecule H_2, with two protons and two electrons, plays a role in the theory of molecular electronic structure similar to that of the hydrogen atom for atomic theory. For the hydrogen molecule we have no exact solution in closed form, though. But fortunately, it is possible by means of fairly simple procedures to construct approximate wave functions which contain some of the fundamental properties that one can expect for an exact solution. Specializing the BO Hamiltonian (1) to this case, we have

$$\mathbf{H} = \frac{1}{R} - \frac{1}{2}[\Delta_1 + \Delta_2]$$
$$- \left[\frac{1}{r_{A1}} + \frac{1}{r_{B1}} + \frac{1}{r_{A2}} + \frac{1}{r_{B2}}\right] + \frac{1}{r_{12}}. \qquad \text{(IV.4)}$$

EXERCISES

8. Draw a figure showing the position of the two protons and the two electrons and identify the distances appearing in (4).

9. Introduce a coordinate system with the z-axis along the molecular axis. Place the origin in the molecular midpoint so that the z-coordinates of the nuclei are $\pm R/2$. Denote the electronic polar coordinates with respect to that origin $\mathbf{r}_1 = (r_1, \vartheta_1, \varphi_1)$ and $\mathbf{r}_2 = (r_2, \vartheta_2, \varphi_2)$. Express the distances r_{A1}, r_{B1}, r_{A2}, r_{B2} between nuclei and electrons in terms of these polar coordinates and R.

10. Express the interelectronic distance r_{12} in terms of **(a)** the vectors $\mathbf{r}_{A1}$ and $\mathbf{r}_{A2}$ and **(b)** the vectors $\mathbf{r}_{B1}$ and $\mathbf{r}_{B2}$.

As we have seen in the atomic case (Section III.B), any wave function for a two-electron system can be factorized in a spin part and a function that depends only on the spatial variables. The ground states of most molecules are singlets. For a two-electron system such as H_2 this means that the spin function is [α_1 is short for $\alpha(\zeta_1)$, etc.]

$$\Theta(\zeta_1, \zeta_2) = \frac{1}{\sqrt{2}} [\alpha_1 \beta_2 - \beta_1 \alpha_2]. \qquad \text{(IV.5)}$$

The total two-electron function is then

$$\Psi(x_1, x_2) = \Phi(\mathbf{r}_1, \mathbf{r}_2)\Theta(\zeta_1, \zeta_2). \qquad \text{(IV.6)}$$

EXERCISES

11. How does the spin function in (5) transform under permutation of its coordinates?

12. Which condition is imposed on the total wave function $\Psi(x_1, x_2)$ in (6) with respect to permutations of its coordinates?

13. What, then, is the relation between the spatial function $\Phi(\mathbf{r}_1, \mathbf{r}_2)$ in (6) and the function $\Phi(\mathbf{r}_2, \mathbf{r}_1)$ obtained by permutation of the coordinates $\mathbf{r}_1$ and $\mathbf{r}_2$?

14. Why does the spin function (5) represent a singlet?

There are basically two different ways of considering molecules with respect to their constituents. One can think of a molecule as consisting of atoms, but then one must be aware of the fact that it is impossible to find any kind of borderline between the atoms in the molecule. The atoms that make up a molecule have independent existence only when the molecule has dissociated into its constituents. This way of looking at molecules has given rise to the valence bond (VB) method, which was used when quantum mechanics was first applied to a chemical problem. For a long time this method has been regarded as obsolete. In recent years a powerful new version of the valence bond method is, however, rapidly becoming an important tool in molecular science. (See the paper by Cooper, Gerratt, and Raimondi referred to at the end of the chapter.)

One can also think of molecules as consisting of nuclei and electrons. The Hamiltonian (1) obviously represents the starting point for such a point of view. But as we will see first for the hydrogen molecule, the main point is to work with *molecular orbitals* instead of atomic orbitals. The quantum chemical expression for this other way of looking at molecules is the molecular orbital (MO) method, which has long dominated the theory of the electronic structure of matter.

Valence Bond Method for H_2

The functional raw material consists of the ground-state hydrogen atom wave functions, one "on" each nucleus:

$$a(1) = \frac{1}{\sqrt{\pi}} e^{-r_{A1}}; \qquad b(1) = \frac{1}{\sqrt{\pi}} e^{-r_{B1}}. \qquad (IV.7)$$

EXERCISES

15. Plot the orbital $a(1)$ as a function of the distance $r_{A1} = |\mathbf{r}_{A1}|$.

16. Use the result of Exercise 9 to plot r_{A1} as a function of R for $0 \leq R < \infty$ for some values of r_1 and ϑ_1. What do the results tell about the orbital $a(1)$ when the molecule dissociates?

17. Show that the orbitals (7) are normalized.

18. What is the meaning of the statement that the orbital $a(1)$ is "situated on" nucleus A?

The orbitals (7) are not orthogonal:

$$S = S(R) = \int dv \, a(1)b(1) = e^{-R}\left(1 + R + \frac{R^2}{3}\right). \quad \text{(IV.8)}$$

EXERCISES

19. The explicit expression for the overlap integral (8) as a function of the internuclear distance can be obtained, for example, with elliptic coordinates:

$$\mu_1 = \frac{1}{R}(r_{A1} + r_{B1});$$

$$\nu_1 = \frac{1}{R}(r_{A1} - r_{B1}); \quad \varphi_1;$$

$$dv_1 = \frac{R^3}{8}(\mu_1^2 - \nu_1^2);$$

$$1 \leq \mu_1 < \infty; \quad -1 \leq \nu_1 \leq 1;$$

$$0 \leq \varphi_1 \leq 2\pi.$$

Carry out these integrations and verify (8).

20. Plot $S = S(R)$ for $0 \leq R < \infty$.

21. Look up the experimental value for the equilibrium internuclear distance in the hydrogen molecule. Express that distance in the atomic unit for length called bohr: 1 bohr = 0.529 Å, and calculate the value of the overlap integral $S(R)$ for the equilibrium distance expressed in bohr.

In the VB method the spatial two-electron function $\Phi(\mathbf{r}_1, \mathbf{r}_2)$ in (6) is constructed as follows:

$$\Phi(\mathbf{r}_1, \mathbf{r}_2) = N_{VB}[a(1)b(2) + b(1)a(2)]. \qquad (IV.9)$$

EXERCISES

22. How does $\Phi(\mathbf{r}_1, \mathbf{r}_2)$ transform under permutations of $\mathbf{r}_1$ and $\mathbf{r}_2$? Is that as it should be?

23. Notice carefully the notation in (9) and (7). Write (9) explicitly as a function of $\mathbf{r}_1$ and $\mathbf{r}_2$.

24. Calculate the normalization constant N_{VB} in (9). Combine the result with the result of a previous exercise to plot N_{VB} as a function R. Is it reasonable that $N_{VB} = 1/\sqrt{2}$ for $R \to \infty$?

The fundamental quantity in a quantum mechanical study of a diatomic molecule is the total energy E as a function of the internuclear distance R: $E = E(R)$. A meaningful wave function should, first, give a lower total energy than that of the free constituents for some equilibrium distance R_e: $E(R_e) < E(\infty)$. It is also desirable that the wave function yield an energy which tends to that of the two free constituents for very large distances: $E(\infty) = 2E_{atom}$.

According to the basic rules of quantum mechanics, the total energy of a hydrogen molecule characterized by a total wave function Ψ is given by

$$E = E(R) = \frac{\langle \Psi | \mathbf{H} | \Psi \rangle}{\langle \Psi | \Psi \rangle}. \qquad \text{(IV.10)}$$

Here $\mathbf{H}$ is the Hamiltonian (4) and the Dirac notation means

$$\langle \Psi | \mathbf{H} | \Psi \rangle = \int dx_1\, dx_2\, \Psi^*(x_1, x_2) \mathbf{H} \Psi(x_1, x_2); \qquad \text{(IV.11}a)$$

$$\langle \Psi | \Psi \rangle = \int dx_1\, dx_2\, \Psi^*(x_1, x_2) \Psi(x_1, x_2). \qquad \text{(IV.11}b)$$

Using a normalized function Ψ and (6), we first simplify (10) to

$$E(R) = \langle \Psi | \mathbf{H} \Psi \rangle = \langle \Phi | \mathbf{H} | \Phi \rangle. \qquad \text{(IV.12)}$$

EXERCISES

25. Describe in words what is done in the last step of (12).

26. Write out $\langle \Phi | \mathbf{H} | \Phi \rangle$ explicitly and check that this quantity is indeed a function of the internuclear distance R only.

The final expression in (12) can be calculated directly, but we can save time and energy by introducing a few tricks that will be equally (or more) useful for other systems. We first notice that (9) can be written

$$\Phi(\mathbf{r}_1, \mathbf{r}_2) = N_{VB}[\mathbf{1} + \mathbf{P}_{12}^r]a(1)b(2). \qquad (IV.13)$$

where $\mathbf{P}_{12}^r$ is the permutation operator that interchanges $\mathbf{r}_1$ and $\mathbf{r}_2$.

EXERCISES

27. Check that the function (9) is symmetric under $\mathbf{P}_{12}^r$:

$$\mathbf{P}_{12}^r\Phi(\mathbf{r}_1, \mathbf{r}_2) = \Phi(\mathbf{r}_2, \mathbf{r}_1) = \Phi(\mathbf{r}_1, \mathbf{r}_2).$$

28. How can the result in Exercise 27 be obtained from (13) using the properties of the permutation operator?

29. Which operator is obtained when the permutation operator is squared: $(\mathbf{P}_{12}^r)^2 = ?$

Apart from a constant factor, the operator in brackets in (13) is a so-called *projection operator* (check!):

$$[\mathbf{1} + \mathbf{P}_{12}^r][\mathbf{1} + \mathbf{P}_{12}^r] = 2[\mathbf{1} + \mathbf{P}_{12}^r]. \qquad (IV.14)$$

A projection operator is thus an operator that does not change when it is squared. The expression (14) should be compared with the properties of the antisymmetrizer (III.33), but one should also carefully notice the differences.

30. Show that the operator $\mathbf{1} + \mathbf{P}_{12}^{r}$ is Hermitian.

31. Show that the operator $\mathbf{1} + \mathbf{P}_{12}^{r}$ commutes with the Hamiltonian (4) for H_2. What does that result imply for the corresponding eigenfunctions?

Using these two properties and (14) together with the turn over rule (cf. section II.H), we can now simplify the calculation of the total energy (12):

$$
\begin{aligned}
E_{VB}(R) &= \langle \Phi | \mathbf{H} | \Phi \rangle \\[2mm]
&= N_{VB}^2 \langle [\mathbf{1} + \mathbf{P}_{12}^{r}] a(1)b(2) | \mathbf{H} | [\mathbf{1} + \mathbf{P}_{12}^{r}] a(1)b(2) \rangle \\[2mm]
&= N_{VB}^2 \langle a(1)b(2) | \mathbf{H} | [\mathbf{1} + \mathbf{P}_{12}^{r}]^2 a(1)b(2) \rangle \\[2mm]
&= 2 N_{VB}^2 \langle a(1)b(2) | \mathbf{H} | [\mathbf{1} + \mathbf{P}_{12}^{r}] a(1)b(2) \rangle \\[2mm]
&= 2 N_{VB}^2 [\langle a(1)b(2) | \mathbf{H} | a(1)b(2) \rangle \\[2mm]
&\quad + \langle a(1)b(2) | \mathbf{H} | b(1)a(2) \rangle].
\end{aligned}
\qquad \text{(IV.15)}
$$

32. Verify all the steps in (15).

33. What do we gain by proceeding this way as compared to a direct calculation with (9) in $\langle \Phi | \mathbf{H} | \Phi \rangle$?

Now we need an explicit expression for the BO Hamiltonian (4), which we write as follows:

$$\mathbf{H} = \mathbf{H}_0 + \mathbf{h}_1 + \mathbf{h}_2 + \frac{1}{r_{12}}. \qquad \text{(IV.16)}$$

Here $\mathbf{H}_0$ is the nuclear repulsion and $\mathbf{h}_i$, $i = 1, 2$, are the one-electron operators.

EXERCISES

34. What is $\mathbf{H}_0$ explicitly? Why is that operator singled out as a separate term?

35. What are $\mathbf{h}_1$ and $\mathbf{h}_2$ explicitly? Notice that $\mathbf{r}_1$ plays the same role in $\mathbf{h}_1$ as $\mathbf{r}_2$ does in $\mathbf{h}_2$.

36. Calculate explicitly, also using the expression for N_{VB} obtained earlier, the final expression for that part of the total energy $E_{\text{VB}}(R)$ which is due to the nuclear repulsion. Is the result what could be expected?

The calculation of that part of $E_{VB}(R)$ which is due to the one-electron operators provides an instructive exercise in the meaning of the term *dummy variable:*

$$\langle a(1)b(2)|\mathbf{h}_1|a(1)b(2)\rangle = \langle a(1)|\mathbf{h}_1|a(1)\rangle \langle b|b\rangle$$

$$= \langle a|\mathbf{h}_1|a\rangle; \qquad \text{(IV.17}a\text{)}$$

$$\langle a(1)b(2)|\mathbf{h}_2|a(1)b(2)\rangle = \langle a|a\rangle \langle b|\mathbf{h}_2|b\rangle$$

$$= \langle b|\mathbf{h}_2|b\rangle$$

$$= \langle b|\mathbf{h}_1|b\rangle$$

$$= \langle a|\mathbf{h}_1|a\rangle; \qquad \text{(IV.17}b\text{)}$$

$$\langle a(1)b(2)|\mathbf{h}_1|b(1)a(2)\rangle = \langle a|\mathbf{h}_1|b\rangle \langle b|a\rangle$$

$$= S\langle a|\mathbf{h}_1|b\rangle; \qquad \text{(IV.17}c\text{)}$$

$$\langle a(1)b(2)|\mathbf{h}_2|b(1)a(2)\rangle = \langle a|b\rangle \langle b|\mathbf{h}_2|a\rangle$$

$$= S\langle b|\mathbf{h}_2|a\rangle$$

$$= S\langle b|\mathbf{h}_1|a\rangle$$

$$= S\langle a|\mathbf{h}_1|b\rangle. \qquad \text{(IV.17}d\text{)}$$

EXERCISES

37. Verify each step in these expressions explicitly. Notice carefully the meaning of the various details in the formulas.

38. Some of the steps in (17) are due to "geometrical" symmetry: both $\mathbf{h}_1$ and $\mathbf{h}_2$ are invariant under exchange of the nuclei A and B. Show that and notice which parts of the four formulas depend on this property.

39. The operators $\mathbf{h}_1$ and $\mathbf{h}_2$ are also Hermitian. Why? Where in (17) is that property used?

The calculation of that part of the total energy which is due to the two-electron operator provides an opportunity to introduce the Mulliken notation for these integrals. We have

$$\left\langle a(1)b(2) \left| \frac{1}{r_{12}} \right| a(1)b(2) \right\rangle$$

$$= \int dv_1 \, dv_2 \, \frac{a(1)b(1)b(2)b(2)}{r_{12}} = (aa|bb); \quad \text{(IV.18a)}$$

$$\left\langle a(1)b(2) \left| \frac{1}{r_{12}} \right| b(1)a(2) \right\rangle$$

$$= \int dv_1 \, dv_2 \, \frac{a(1)b(1)b(2)a(2)}{r_{12}} = (ab|ba). \quad \text{(IV.18b)}$$

The Mulliken notation in these expressions, $(aa|bb)$ and $(ab|ba)$, means that we interpret the integrals as electrostatic interaction energies between the charge distributions $a(1)a(1)$ and $b(2)b(2)$ in (18a), and between $a(1)b(1)$ and $b(2)a(2)$ in (18b).

EXERCISE

40. Collect the various contributions to the total energy (15) and express it in a minimal number of quantities, namely S and

$$J' = \int dv_1 \, \frac{a(1)a(1)}{r_{B1}}; \quad J = (aa|bb); \quad \text{(IV.19a)}$$

$$K' = \int dv_1 \, \frac{a(1)b(1)}{r_{A1}}; \quad K = (ab|ab). \quad \text{(IV.19b)}$$

Like $S = S(R)$, these integrals can be calculated explicitly as functions of R. The net result is a function $E_{VB}(R)$—a potential energy curve—with the following features. For very large values of the internuclear distance, $E_{VB}(R)$ tends to the energy of the two free hydrogen atoms. When R decreases, the potential energy decreases to a minimum at the equilibrium distance R_e. For even smaller distances it increases indefinitely. The binding energy E_b is the difference $E_{VB}(\infty) - E_{VB}(R_e)$.

EXERCISES

41. The binding energy E_b is the gain in energy that is obtained when the molecule is formed from two atoms with ground-state energies E_{1s}. Express E_b in terms of E_{1s} and the quantities (19).

42. Which property of the function $E_{VB}(R)$ determines the equilibrium distance R_e?

43. The derivative of $E_{VB}(R)$ with respect to R is positive to the right of R_e and negative on the other side. Try to interpret the result physically (or chemically).

44. Draw a potential energy curve with the properties indicated. Identify the various features. Compare the result with, for example, Atkins' book.

The fact that $E_{VB}(R)$ provides a meaningful expression for the total energy as a function of the internuclear distance does not mean that (9) is an exact solution of the Schrödinger equation for H_2. But it does mean that (9) represents a qualitatively correct picture. The quantitative agreement between experimental and theoretical values is also impressive:

$$E_b^{VB} = 303 \text{ kJ/mol} = 3.14 \text{ eV} = 0.115 \text{ hartree;}$$

$$E_b^{Exp} = 457 \text{ kJ/mol} = 4.73 \text{ eV} = 0.174 \text{ hartree;}$$

$$R_e^{VB} = 80 \text{ pm} = 0.80 \text{ Å} = 1.51 \text{ bohr;}$$

$$R_e^{Exp} = 74 \text{ pm} = 0.74 \text{ Å} = 1.40 \text{ bohr.}$$

Heitler and London first carried out this VB calculation for H_2 in 1926. Their work constitutes the beginning of a new scientific discipline—quantum chemistry.

Molecular Orbital Method for H_2

The molecular orbital (MO) method is inspired by the situation in atoms. The states of the hydrogen atom are known exactly. They are characterized by the atomic orbital (AO), which is occupied by the electron: $1s$, $2s$, $3s$, . . . , $2p$, $3p$, $4p$, . . . , $3d$, $4d$, $5d$, . . . , $4f$, $5f$, $6f$, and so on. In atoms with more than one electron the *type* of AO known from the hydrogen atom can be used to construct approximate solutions, even though no exact solutions are known. That is the background for describing the ground states of, for example, the Li and the C atom as $1s^2 2s$ and $1s^2 2s^2 2p^2$, respectively. In the MO method for molecules we also construct—in one way or another—a set of orbitals, which are filled with the available electrons, usually two per orbital, one for each spin. In the notation for the AOs the letters s, p, d, . . . play the role of symmetry labels. For molecules we also need symmetry labels, but they will be different for different types of molecules. In the case of diatomic molecules we have cylindrical symmetry as distinguished from the spherical symmetry of atoms. Molecules with more than two nuclei have symmetries characterized by the point groups. Symmetry considerations play an essential part in all discussions of electronic structure, but we will not pursue these aspects here. Instead, we go directly to the simplest kind of MO function for the hydrogen molecule. The total function is still of the form (6), but instead of (9) we have a spatial function that is a product of two so-far-unknown orbitals:

$$\Phi_{MO}(\mathbf{r}_1, \mathbf{r}_2) = \psi(\mathbf{r}_1)\psi(\mathbf{r}_2). \qquad \text{(IV.20)}$$

83

45. How does (20) transform under permutation of its variables? Does it behave as it should?

46. What is the normalization constant of the two-electron function $\Phi_{MO}(\mathbf{r}_1, \mathbf{r}_2)$ if the orbital $\psi(\mathbf{r})$ itself is assumed to be normalized?

Thus $\Phi_{MO}(\mathbf{r}_1, \mathbf{r}_2)$ is a product of two identical orbitals that are functions of different variables. When (20) is combined with a spin function as in (5), we see explicitly what is meant by saying that the orbital ψ is doubly occupied, one with α and one with β spin.

47. Write (6) with (20) as a determinant. That provides an even better illustration of the previous sentence.

48. Expand that determinant so that a linear combination of products of spin orbitals is obtained. Verify that the result agrees with what is obtained from (5) with (20) by explicit multiplication.

To calculate the total MO energy, we proceed as in the VB case [cf. (10)–(12)]:

$$E_{MO}(R) = \frac{\langle \Psi_{MO} | \mathbf{H} | \Psi_{MO} \rangle}{\langle \Psi_{MO} | \Psi_{MO} \rangle} = \langle \Phi_{MO} | \Phi_{MO} \rangle. \quad \text{(IV.21)}$$

EXERCISES

49. Use (20) in (21) in order to express E_{MO} in terms of the orbital ψ.

50. Does the result of Exercise 49 contain any term that appears also in the valence bond energy?

51. Notice that the result of Exercise 49 means that E_{MO} can be regarded as a *functional* of ψ: for any choice of that function we get a corresponding value of the energy.

In order to go further we need more explicit information about the orbital ψ. This can be obtained using the variational principle. We will not pursue that path here, however. Instead, we make an explicit ansatz:

$$\psi(\mathbf{r}_1) = N_{MO}[a(1) + b(1)], \quad \text{(IV.22)}$$

where $a(1)$ and $b(1)$ are the atomic orbitals (7). This means that the MO has been approximated as a *Linear Combination of Atomic Orbitals* (*LCAO*). That aproximation—MO-LCAO—is actually the most common way of using the MO method in practice. Its intuitive background is that for very large internuclear distances we can expect the system to consist of noninteracting atoms. For small distances the AOs interfere with each other, and as we will see, an ansatz like (22) is often enough to produce a reasonable first approximation to the MOs.

52. Determine the normalization constant N_{MO} in (22) so that the MO ψ is normalized when [cf. (7)] the AOs a and b are themselves normalized. Notice that these AOs are not orthogonal.

53. Plot $N_{MO} = N_{MO}(R)$ as a function of the internuclear distance. Is its value $1/\sqrt{2}$ for $R \to \infty$ reasonable?

54. In (22) the AO $a(1)$ has the same coefficient as the other AO $b(1)$. Try to find reasons for that choice.

By substituting (22) in the expression [cf. Exercise 49]

$$E_{MO}(R) = \mathbf{H}_0 + 2\langle\psi|\mathbf{h}_1|\psi\rangle + (\psi\psi|\psi\psi), \quad \text{(IV.23)}$$

we can express the total MO energy in the same types of quantities as were used in the VB case:

$$E_{MO}(R) = \frac{1}{R} + 2E_{1s} - \frac{2(J' + K')}{1 + S} + \frac{1}{2(1 + S)^2}$$

$$\cdot [2K + J + (aa|aa) + 4(aa|ab)]. \quad \text{(IV.24)}$$

55. Verify all the steps in (24). What is E_{1s}?

56. Calculate the difference $E_{MO}(R) - E_{VB}(R)$.

57. For $R \to \infty$ all integrals vanish that contain both a and b. (Why?) Use that fact together with $(aa|aa) = \frac{5}{8}$ to show that for $R \to \infty$.

$$E_{MO}(R) \to 2E_{1s} + \frac{5}{16}.$$

Is that as it should?

Near the equilibrium $E_{MO}(R)$ behaves more reasonably:

$$E_b^{MO} = 350 \text{ kJ/mol} = 3.63 \text{ eV} = 0.133 \text{ hartree};$$

$$R_e^{MO} = 74 \text{ pm} = 0.74 \text{ Å} = 1.40 \text{ bohr}.$$

Here the binding energy has been calculated as the difference between the exact energy of two noninteracting hydrogen atoms and the value of $E_{MO}(R)$ at its minimum. As should be clear from Exercise 57, the value of $E_{MO}(R)$ for $R \to \infty$ does not represent the energy of two neutral atoms. This is a fundamental problem with the MO method, which we will not pursue here. We recall, however, that $E_{VB}(R) \to 2E_{1s}$ for $R \to \infty$, as it should. Even though the valence bond method thus performs better than the MO method for H_2, it is the latter that has dominated the development for over 60 years. To a large extent this is due to the fact that it is much easier to apply the MO method to larger systems than H_2. It is also possible to go beyond the MO method and introduce improvements in a systematic fashion. In recent years the VB method has, however, experienced somewhat of a renaissance. (See the paper by Cooper, Gerratt, and Raimondi referred to at the end of the chapter.)

It is instructive to compare the two spatial functions (9) and (20) for $R \to \infty$. We get

$$\Phi_{VB}(\mathbf{r}_1, \mathbf{r}_2) \to \frac{1}{\sqrt{2}} [a(1)b(2) + b(1)a(2)]; \qquad \text{(IV.25a)}$$

$$\Phi_{MO}(\mathbf{r}_1, \mathbf{r}_2) \rightarrow \frac{1}{\sqrt{2}} \left\{ \frac{1}{\sqrt{2}} \left[a(1)b(2) + b(1)a(2) \right] \right.$$

$$\left. + \frac{1}{\sqrt{2}} \left[a(1)a(2) + b(1)b(2) \right] \right\}. \quad (IV.25b)$$

EXERCISES

58. Verify all the details in (25).

59. Are the two functions in (25) normalized?

60. Try to interpret the square brackets in (25)—remember that there is no interaction between electrons on different nuclei when $R \rightarrow \infty$.

In (25b) we see that the MO function for $R \rightarrow \infty$ is a linear combination of the VB function for $R \rightarrow \infty$, and an *ionic* function,

$$\frac{1}{\sqrt{2}} \left[a(1)a(2) + b(1)b(2) \right]. \quad (IV.26)$$

The interpretation of this is that in the MO method there is a certain nonzero probability (how large?) for both electrons to be found at the same nucleus. That raises the total energy above the correct dissociation energy $2E_{1s}$.

The integrals appearing in the total energies of the VB and MO methods can be calculated explicitly. They are given by

(see, e.g., the book by Slater referred to at the end of the chapter)

$$J' = \frac{1}{R}[1 - e^{-2R}(1 + R)]; \qquad \text{(IV.27a)}$$

$$K' = e^{-R}(1 + R); \qquad \text{(IV.27b)}$$

$$J = \frac{1}{R} - \frac{e^{-2R}}{2}\left[\frac{2}{R} + \frac{11}{4} + \frac{3R}{2} + \frac{R^2}{3}\right]; \qquad \text{(IV.27c)}$$

$$K = \frac{1}{5}\left\{\frac{6}{R}[S^2(\gamma + \ln R) - S(-R)^2 \, \mathrm{Ei}(-4R)\right.$$

$$+ 2SS(-R) \, \mathrm{Ei}(-2R)]$$

$$\left. - e^{-2R}\left[-\frac{25}{8} + \frac{23R}{4} + 3R^2 + \frac{R^3}{3}\right]\right\}; \qquad \text{(IV.27d)}$$

$$(aa|ab) = \frac{1}{2}\left\{e^{-R}\left[2R + \frac{1}{4} + \frac{5}{8R}\right]\right.$$

$$\left. - e^{-3R}\left[\frac{1}{4} + \frac{5}{8R}\right]\right\}. \qquad \text{(IV.27e)}$$

Here γ is Euler's constant ($\gamma = 0.578722 \ldots$) and the function $\mathrm{Ei}(x)$ is the exponential integral. $S = S(R)$ is the overlap integral given in (8).

EXERCISES

61. Plot the quantities (27) as functions of R.

62. Use (27) to calculate and plot $E_{VB}(R)$.

63. Use (27) to calculate and plot $E_{MO}(R)$.

Bonding and Antibonding Orbitals

So far we have only used the molecular orbital (22) to describe the ground state of the hydrogen molecule. We have not given any reasons for choosing the same coefficients for the two AOs a and b in (22). Intuitively, one can expect from the symmetry of a homonuclear diatomic molecule like H_2 that the two AOs should appear in a symmetric way. But nothing prevents us from letting them have coefficients with different signs. We write (22) and its counterpart with a negative sign as

$$\psi_g(\mathbf{r}_1) = \frac{1}{\sqrt{2(1 + S)}} [a(1) + b(1)]; \quad (IV.28a)$$

$$\psi_u(\mathbf{r}_1) = \frac{1}{\sqrt{2(1 - S)}} [a(1) - b(1)]. \quad (IV.28b)$$

---

64. Verify that $\psi_u(\mathbf{r}_1)$ is normalized.

65. Calculate the overlap integral of the two MOs in (28).

66. The letters g and u refer to the German words *gerade (even)* and *ungerade (odd)*, respectively. The two MOs are even and odd, respectively, with respect to a certain symmetry operation. Which one?

The two MOs (28) can be used to build up two-electron functions for different states of the molecule. If we limit ourselves to singlets, the spatial function must be symmetric under exchange of $\mathbf{r}_1$ and $\mathbf{r}_2$. For the ground state we have used

$$\Phi_0(\mathbf{r}_1, \mathbf{r}_2) = \psi_g(\mathbf{r}_1)\psi_g(\mathbf{r}_2). \qquad \text{(IV.29)}$$

The function

$$\Phi_1(\mathbf{r}_1, \mathbf{r}_2) = \psi_u(\mathbf{r}_1)\psi_u(\mathbf{r}_2) \qquad \text{(IV.30)}$$

is also symmetric, however, and so corresponds to a possible spatial component for a singlet.

EXERCISES

67. Are the functions (29) and (30) normalized?

68. Calculate the overlap integral between $\Phi_0(\mathbf{r}_1, \mathbf{r}_2)$ and $\Phi_1(\mathbf{r}_1, \mathbf{r}_2)$.

We now have all the machinery necessary for calculating the total energy of the hydrogen molecule in a state with the spatial function (30):

$$E_1(R) = \langle \Phi_1 | \mathbf{H} | \Phi_1 \rangle. \qquad \text{(IV.31)}$$

EXERCISES

69. Proceed as with the ground-state energy to calculate (31) in terms of the quantities (27).

70. Plot $E_1(R)$ as a function of R. Compare with $E_0(R)$.

71. Calculate $E_1(R)$ for $R \to \infty$.

A triplet function for a two-electron system can also be written in the form

$$\Psi_t(x_1, x_2) = \Phi(\mathbf{r}_1, \mathbf{r}_2)\Theta_t(\zeta_1, \zeta_2), \qquad \text{(IV.32)}$$

but with the spin function chosen as one of the following three triplet functions:

$$\alpha_1\alpha_2, \qquad \frac{1}{\sqrt{2}}[\alpha_1\beta_2 + \beta_1\alpha_2], \qquad \beta_1\beta_2. \qquad \text{(IV.33)}$$

EXERCISES

72. How do the three spin functions (33) transform under permutations of their arguments?

73. What condition is imposed on the total function (32) with respect to permutations of x_1 and x_2?

74. What does this imply for the spatial function in (32)?

With the two orbitals (28), there is only one possibility for $\Phi(\mathbf{r}_1, \mathbf{r}_2)$ in (32), namely

$$\Phi(\mathbf{r}_1, \mathbf{r}_2) = \frac{1}{\sqrt{2}} [\psi_g(\mathbf{r}_1)\psi_u(\mathbf{r}_2) - \psi_u(\mathbf{r}_1)\psi_g(\mathbf{r}_2)]. \quad \text{(IV.34)}$$

EXERCISES

75. Calculate the overlap integrals between a spin function of type (5) and the three functions (33).

76. Is the spatial two-electron function (34) normalized?

77. Express (34) in terms of the atomic orbitals a and b. Is there any relation to the valence bond function?

78. Calculate the total energy in the state represented by (34) for $R \to \infty$. Does it "behave properly"?

The three states discussed here can be characterized with the notations (why?)

$$(g\alpha, g\beta), \quad (u\alpha, u\beta), \quad (g\alpha, u\alpha). \quad \text{(IV.35)}$$

The ground-state function represents a state with a minimum in the potential curve for a certain equilibrium distance R_e. When one of the electrons is "promoted" to the spin orbital $u\alpha$, we get a triplet state with a repulsive potential curve (i.e., without any minimum). If both electrons are placed in the u orbital with different spins, we get an excited singlet state with a potential curve, which lies even higher. This is the background for calling ψ_g a bonding and ψ_u an antibonding orbital.

Momentum-Space Functions for the Hydrogen Molecule

It is instructive to investigate the meaning of momentum space in connection with the simple functions we have used here for the hydrogen molecule. Then it is helpful to notice that

$$r_{A1} = \left| \mathbf{r}_1 + \frac{\mathbf{R}}{2} \right|; \qquad r_{B1} = \left| \mathbf{r}_1 - \frac{\mathbf{R}}{2} \right|; \qquad \text{(IV.36)}$$

From (III.10) we have for a $1s$ function,

$$\underline{\phi}(\mathbf{p}) = \frac{\sqrt{2}}{\pi p} \int_0^\infty re^{-r} \sin(pr)\, dr = \frac{1}{\sqrt{4\pi}} \underline{R}_{1s}(p). \qquad \text{(IV.37)}$$

EXERCISES

79. Verify (37). Which steps have been carried out in the basic transformation formula (III.5)?

80. Calculate (37) explicitly, using the integral

$$\int_0^\infty re^{-r} \sin(pr)\, dr = 2p/(1 + p^2)^2.$$

81. Plot the function $\underline{R}_{1s}(p)$.

If a function ϕ in position space is centered at another point $\mathbf{m}$, the corresponding momentum-space function gets multiplied by a phase factor determined by $\mathbf{m}$:

$$\frac{1}{\sqrt{8\pi^3}} \int dv \, \phi(\mathbf{r} - \mathbf{m}) e^{-i\mathbf{p}\cdot\mathbf{r}}$$

$$= \frac{1}{\sqrt{8\pi^3}} \int dv' \, \phi(\mathbf{r}') e^{-i\mathbf{p}\cdot(\mathbf{r}' + \mathbf{m})}$$

$$= \underline{\phi}(\mathbf{p}) e^{-i\mathbf{p}\cdot\mathbf{m}}. \qquad\qquad \text{(IV.38)}$$

EXERCISES

82. Use (38) to calculate the momentum-space functions corresponding to a and b in (7).

83. Calculate the momentum-space functions corresponding to the bonding and antibonding orbitals (28). Even though the function $\underline{\phi}_{1s}(\mathbf{p})$ never vanishes, $\underline{\psi}_g(\mathbf{p})$ and $\underline{\psi}_u(\mathbf{p})$ can vanish. How is that possible? For which values of $\mathbf{p}$ does that happen?

84. In what way do $\underline{\psi}_g(\mathbf{p})$ and $\underline{\psi}_u(\mathbf{p})$ depend on the direction of $\mathbf{p}$?

85. What are the values of $\underline{\psi}_g(\mathbf{p})$ and $\underline{\psi}_u(\mathbf{p})$ for $p = 0$?

86. How do $\underline{\psi}_g(\mathbf{p})$ and $\underline{\psi}_u(\mathbf{p})$ vary with R?

Densities for the Hydrogen Molecule

For a two-electron function of type (6) with a symmetric spatial function we get with (III.58) and (III.59) the following density:

$$\rho(\mathbf{r}) = \int dx_1 \, dx_2 \, \Psi^*(x_1, x_2) \sum_{i=1}^{2} \delta(\mathbf{r} - \mathbf{r}_i) \Psi(x_1, x_2)$$

$$= \int dv_1 \, dv_2 \, \Phi^*(\mathbf{r}_1, \mathbf{r}_2) \sum_{i=1}^{2} \delta(\mathbf{r} - \mathbf{r}_i) \Phi(\mathbf{r}_1, \mathbf{r}_2)$$

$$= 2 \int dv_1 \, |\Phi(\mathbf{r}_1, \mathbf{r})|^2. \qquad \text{(IV.39)}$$

EXERCISES

87. What is the origin of the factor 2 in the last line of (39)?

88. Where in (39) is the property $\Phi(\mathbf{r}_1, \mathbf{r}_2) = \Phi(\mathbf{r}_2, \mathbf{r}_1)$ used?

The expression (39) can now be used for either the VB or the MO function. In the MO case (29) we get

$$\rho(\mathbf{r}) = 2\psi_g(\mathbf{r})^2. \qquad \text{(IV.40)}$$

89. Use one of Slater's rules for determinants to derive (40) in an alternative way.

90. Express the density in the hydrogen molecule in terms of the AOs a and b when the MO is expressed as a LCAO [cf. (28)].

91. To which density does the ground-state MO density tend for $R \to \infty$?

92. Calculate the integral $\int dv\, \rho_{\mathrm{MO}}(\mathbf{r})$ to check that there are two electrons in the system.

In the VB case we get, with (39) and (9),

$$\rho_{\mathrm{VB}}(\mathbf{r}) = \frac{1}{1 + S^2}\, [b^2(\mathbf{r}) + 2Sa(\mathbf{r})b(\mathbf{r}) + a^2(\mathbf{r})]. \quad \text{(IV.41)}$$

EXERCISES

93. Verify (41). What does $a(\mathbf{r})$ and $b(\mathbf{r})$ actually mean? In other words, how can these functions be calculated explicitly at a given point $\mathbf{r}$?

94. Can (41) be calculated with any of Slater's rules for determinants?

95. Calculate $\int dv \, \rho_{VB}(\mathbf{r})$ to check that there are two electrons in the system.

96. To which density does (41) tend for $R \to \infty$?

___

Momentum Distribution

The momentum distribution in the MO case is [cf. (III.18)]

$$\underline{\rho}_{MO}(\mathbf{p}) = 2|\underline{\psi}_g(\mathbf{p})|^2. \qquad (IV.42)$$

Using the result of Exercise 83, we see that this can be written

$$\underline{\rho}_{MO}(\mathbf{p}) = \frac{4}{1+S} |\underline{\phi}_{1s}(\mathbf{p})|^2 \cos^2\left(\frac{\mathbf{p} \cdot \mathbf{R}}{2}\right). \qquad (IV.43)$$

___

EXERCISES

97. How does this momentum distribution depend on the direction of $\mathbf{p}$?

98. How does $\underline{\rho}_{MO}(\mathbf{p})$ vary with R?

99. What is $\underline{\rho}_{MO}(\mathbf{0})$? Is that what one could expect?

In order to calculate the momentum distribution in the VB case we need a quantity that is a generalization of the density (41), namely the first-order density matrix,

$$\gamma_{VB}(\mathbf{r}, \mathbf{r}') = \frac{1}{1 + S^2} [a(\mathbf{r})a(\mathbf{r}') + Sa(\mathbf{r})b(\mathbf{r}')$$

$$+ Sb(\mathbf{r})a(\mathbf{r}') + b(\mathbf{r})b(\mathbf{r}')]. \qquad \text{(IV.44)}$$

EXERCISES

100. What is $\gamma_{VB}(\mathbf{r}, \mathbf{r})$—that is, when the two variables in (44) are equal?

101. Is there any relation between $\gamma_{VB}(\mathbf{r}, \mathbf{r}')$ and $\gamma_{VB}(\mathbf{r}', \mathbf{r})$?

102. The quantity (44) can be written in the form

$$[a(\mathbf{r}) \quad b(\mathbf{r})]\Gamma\begin{bmatrix} a(\mathbf{r}') \\ b(\mathbf{r}') \end{bmatrix},$$

where Γ is a 2×2 matrix. Calculate the elements in the matrix.

The momentum distribution in the VB case is now obtained from [cf. (III.19)] the expression

$$\underline{\rho}_{VB}(\mathbf{p}) = \frac{1}{8\pi^3} \int dv \, dv' \, \gamma_{VB}(\mathbf{r}, \mathbf{r}') e^{i\mathbf{p}\cdot(\mathbf{r}' - \mathbf{r})}. \qquad (IV.45)$$

EXERCISES

103. Use the result of Exercise 82 to write $\rho_{VB}(\mathbf{p})$ in terms of the function $\underline{\phi}_{1s}(\mathbf{p})$ (which function is that?)

104. Compare the result with the MO case, particularly for $R \to \infty$.

105. Calculate $\int d\mathbf{p} \, \underline{\rho}_{VB}(\mathbf{p})$ from (45). Is the result what one should expect?

Form Factors

We recall from (III.20) that the form factor is the Fourier transform of the density:

$$F(\mathbf{p}) = \int dv \, \rho(\mathbf{r}) e^{i\mathbf{p}\cdot\mathbf{r}}. \qquad (IV.46)$$

This should be carefully distinguished from the momentum-space function, which is the Fourier transform of the position-space function. To calculate the form factor we need the Fourier transform of a *product of orbitals*. It is instructive first to express the Fourier transform of a product of position-space orbitals in

100

terms of their momentum-space counterparts. We have, from (III.5) and (III.7),

$$\underline{\phi}(\mathbf{p}) = \frac{1}{\sqrt{8\pi^3}} \int dv \, \phi(\mathbf{r}) e^{-i\mathbf{p}\cdot\mathbf{r}}; \qquad (IV.47a)$$

$$\phi(\mathbf{r}) = \frac{1}{\sqrt{8\pi^3}} \int d\mathbf{p} \, \underline{\phi}(\mathbf{p}) e^{i\mathbf{p}\cdot\mathbf{r}}. \qquad (IV.47b)$$

The Fourier transform of the product $\phi(\mathbf{r})\psi^*(\mathbf{r})$ is

$$\int dv \, \phi(\mathbf{r})\psi^*(\mathbf{r}) e^{i\mathbf{p}\cdot\mathbf{r}}$$

$$= \frac{1}{8\pi^3} \int d\mathbf{p}' \, d\mathbf{p}'' \, \underline{\phi}(\mathbf{p}')\underline{\psi}^*(-\mathbf{p}'') \int dv \, e^{i(\mathbf{p}+\mathbf{p}'+\mathbf{p}'')\cdot\mathbf{r}}$$

$$= \int d\mathbf{p}' \, d\mathbf{p}'' \, \underline{\phi}(\mathbf{p}')\underline{\psi}^*(-\mathbf{p}'')\delta(\mathbf{p}+\mathbf{p}'+\mathbf{p}'')$$

$$= \int d\mathbf{p}' \underline{\phi}^*(\mathbf{p}')\underline{\psi}^*(\mathbf{p}'+\mathbf{p}). \qquad (IV.48)$$

In words, this can be described by saying that the Fourier transform of the product of two orbitals is the autocorrelation function of their momentum-space counterparts.

EXERCISES

106. Why does the argument of the function ψ have a negative sign in the second line of (48)?

107. Use the result of Exercise 82 to calculate the Fourier transform of the following products: $a(\mathbf{r})a(\mathbf{r})$, $a(\mathbf{r})b(\mathbf{r})$, and $b(\mathbf{r})b(\mathbf{r})$.

108. How do the results of Exercise 107 depend on the vector $\mathbf{R}$?

Using (46) and (48), we get the form factor in the MO case for the hydrogen molecule

$$F_{MO}(\mathbf{p}) = 2 \int d\mathbf{p}' \, \underline{\psi}_g(\mathbf{p}')\underline{\psi}_g(\mathbf{p}' + \mathbf{p}). \qquad \text{(IV.49)}$$

EXERCISES

109. Verify (49).

110. What is $F_{MO}(\mathbf{0})$? How can the result be interpreted?

111. Express (49) in terms of integrals like (48) over the AOs a and b.

112. Calculate the form factor in the VB case and express the result in terms of integrals like (48) over the AOs a and b.

Reciprocal Form Factors

Using (42), we get, with (III.22),

$$B_{MO}(\mathbf{r}) = 2 \int dv' \, \psi_g(\mathbf{r}')\psi_g(\mathbf{r}' + \mathbf{r}). \qquad \text{(IV.50)}$$

EXERCISES

113. Verify (50).

114. Calculate the correlation functions $\int dv'\, a(\mathbf{r}')a(\mathbf{r}' - \mathbf{r})$, $\int dv'\, a(\mathbf{r}')b(\mathbf{r}' - \mathbf{r})$, and $\int dv'\, b(\mathbf{r}')b(\mathbf{r}' - \mathbf{r})$.

115. What are the values of the functions calculated in Exercise 114 for $\mathbf{r} = 0$? Does that agree with what one obtains directly from a definition of the functions?

IV.B. HÜCKEL THEORY FOR AROMATIC MOLECULES

In previous sections on the hydrogen molecule we have shown how quantum mechanics "explains" the covalent chemical bond. Two different approximate wave functions were used as trial functions and the corresponding total energies could be calculated explicitly in closed form as functions of the internuclear distance. In the present section we use quantum mechanics in a rather different way, where the *structure* of the theory is more important than the actual computation of all quantities involved. We describe a much more approximate procedure than the "ab initio" calculations for H_2 but one that has the great merit of being applicable to considerably larger molecules. The organic molecules we treat in the present section are still small, "finite" systems, but the methods we are going to introduce will also turn out to be important for extended systems.

We apply an approximate version of molecular orbital theory to systems consisting of carbon and hydrogen atoms, primarily molecules with "more than single" but "less than double" bonds between the carbon atoms. The main common feature of these molecules is their geometrical structure: all the carbon atoms lie in the same plane. A carbon atom has two core electrons of type $1s$ and four valence electrons. In the free atom these are denoted $2s^2 2p^2$. In a molecule there is no spherical symmetry and the notation s and p becomes meaningless.

EXERCISES

116. In what way is the notation s and p related to the spherical symmetry of a free atom?

117. Why is it possible to speak about the core electrons as $1s$ electrons both in atoms and molecules?

In the planar molecules we are going to work with here, three of the four valence orbitals of each carbon atom form so-called hybrid orbitals with lobes pointing in directions forming 120° with each other and "lying" in the same plane. The planar system formed in this way is called the σ-skeleton of the molecule. Just as the letter s is used for atoms to denote invariance under all rotations around one point, the letter σ is used in these planar molecules to denote functions that are invariant under reflections in the plane.

The hybrid orbitals that make up the σ-skeleton consist of linear combinations of one $2s$, one $2p_x$, and one $2p_y$ orbital on each carbon atom. That apparently implies that we have introduced a coordinate system with the molecular plane as the xy-plane. There is also a $2p_z$ orbital on each atom with a positive lobe above and a negative lobe below the plane. Such a $2p_z$ orbital apparently changes sign under reflection in the plane. That transformation property is denoted π in the present case. Our entire interest in this section is going to be focused on the electrons in these π-orbitals. In other words, both the core electrons and the electrons in the σ-skeleton are going to be regarded as "frozen." We are going to study the structure of the π-electron system in such a frozen background.

Ethylene

Ethene (or ethylene) is the simplest of all π-bonded systems. That molecule is traditionally represented by the formula

$$\text{H} \diagdown \diagup \text{H} \atop \text{H} \diagup \text{C}=\text{C} \diagdown \text{H}$$

which is not only confirmed but also "explained" by quantum mechanical calculations. We recognize the 120° angles on each carbon atom due to the σ-skeleton. The supplementary bond between the carbon atoms is due to two electrons with different spins in a π-orbital.

EXERCISES

118. What is the total number of electrons in the ethene molecule?

119. How can one describe (qualitatively) the bonds between a carbon atom and the neighboring hydrogen atoms?

120. Draw pictures of **(a)** the σ-skeleton in the molecular plane, **(b)** the molecule seen in the positive x-direction, and **(c)** the molecule seen in the positive y-direction. The point is to show the orbital lobes with their sign, if any.

In order to proceed we assume the existence of an effective one-electron Hamiltonian for the π-electrons. Without going into detail as to the background for such an assumption, it can safely be claimed that it is backed up by solid arguments. More important here is to grasp the very idea of an effective Hamiltonian. Like all Hamiltonians it should consist of a kinetic and a potential energy operator. The kinetic part is always the same— the Laplacian operator. We are not going to use this Hamiltonian for detailed calculations, so we will not need any explicit details about the potential part. But it is important to understand that this potential energy operator should represent the combined influence of both the frozen σ-skeleton and of the other π-electrons. It should be added that even though approximate effective Hamiltonians of such a type can indeed be constructed, it is not possible to reduce the exact many-electron problem to such a construction.

The explicit starting point for treatment of the π-electron system in ethene is then an effective one-electron Schrödinger equation,

$$\mathbf{h}\psi = \epsilon\psi. \qquad \text{(IV.51)}$$

The very starting point is thus of a more approximate nature than in the H_2 case. In order to construct approximate solutions of (51) we are, however, going to use the same procedure as in the MO-LCAO case for H_2. In the present case we have at our disposal two $2p_z$-orbitals, one on each carbon atom. For these we introduce the notation ϕ_A and ϕ_B. It is essential in the present procedure that even though we are not going to perform detailed calculations with these functions, we must clearly understand the meaning of the notation.

EXERCISES

121. Look up the explicit expression for a normalized $2p_z$-orbital centered at the origin of a coordinate system. Check the normalization.

122. Introduce a coordinate system with the xy-plane as the plane of the ethene molecule and with the x-axis along the $C-C$ bond. Specify the coordinates for all the six nuclei in terms of the distances R_{CC} and R_{CH}.

123. Give explicit expressions for the two $2p_z$-orbitals ϕ_A and ϕ_B centered at the two carbon nuclei.

The point with the MO-LCAO method is just to approximate the molecular orbitals as linear combinations of certain atomic orbitals. There are both chemical and mathematical reasons for such a procedure. We thus make the ansatz,

$$\psi = \phi_A c_A + \phi_B c_B. \qquad \text{(IV.52)}$$

Here the *basis functions* ϕ_A and ϕ_B are assumed to be known and the coefficients c_A and c_B are so far unknown.

EXERCISES

124. In the H_2 case we used similar LCAOs for both the bonding and the antibonding orbital. Which values do the coefficients have in those cases?

125. In the H_2 case the ratio between the coefficients was taken for granted. Can any particular reason be mobilized for that? Can similar reasoning be used for ethene?

We substitute (52) in (51):

$$\mathbf{h}\phi_A c_A + \mathbf{h}\phi_B c_B = \epsilon\phi_A c_A + \epsilon\phi_B c_B. \qquad \text{(IV.53)}$$

Then we multiply this equation by ϕ_A^* and integrate:

$$h_{AA}c_A + h_{AB}c_B = \epsilon c_A + \epsilon S c_B. \qquad \text{(IV.54)}$$

Here we have apparently introduced the notation

$$h_{AA} = \int dv\, \phi_A^* \mathbf{h}\phi_A; \qquad h_{AB} = \int dv\, \phi_A^* \mathbf{h}\phi_B; \qquad \text{(IV.55}a\text{)}$$

$$S = \int dv\, \phi_A^* \phi_B. \qquad \text{(IV.55}b\text{)}$$

The overlap integral S can be calculated explicitly as in the H_2 case (but notice that the atomic orbitals involved are completely different). For several reasons we are not going to do that. The other integrals in (55) cannot be calculated since we have no explicit expression for the effective Hamiltonian $\mathbf{h}$. We can as well multiply (53) by ϕ_B^* and then integrate. That gives

$$h_{BA}c_A + h_{BB}c_B = \epsilon S c_A + \epsilon c_B. \qquad \text{(IV.56)}$$

EXERCISES

126. Why is there a term ϵc_B in (56) and a term ϵc_A in (54)?

127. In $h_{BA} = h_{AB}$? What condition must be imposed on the effective Hamiltonian $\mathbf{h}$ in order to have that result?

128. Is $h_{AA} = h_{BB}$?

The steps performed from (52) to (54) and (56) mean that the integrodifferential equation (51) has been reduced to a set of two linear equations with the two unknowns c_A and c_B. The problem is now that the coefficients in equations (54) and (56) are themselves unknown.

EXERCISES

129. Which are these coefficients? Notice that the term *coefficient* can refer to different entities, depending on the context.

130. Write explicitly the coefficients for c_A and c_B in the two equations (54) and (56).

In order to go on one makes—following Hückel, who introduced this method at the beginning of the 1930s—three seemingly drastic assumptions:

1. All overlap integrals are neglected: $S = 0$.
2. All diagonal matrix elements of **h** are set equal: $h_{AA} = h_{BB} = \alpha$.
3. All nondiagonal elements of **h** are neglected except those connecting nearest neighbors: $h_{AB} = \beta$.

Of these assumptions, number 1 is easiest to motivate, despite its seemingly obvious absurdity. Assumption 2 is perfectly reasonable since we deal with only one kind of atom. Assumption 3 is also relatively reasonable.

EXERCISES

131. The notation used in (54) and (56) means that we represent the operator **h** by a matrix

$$\mathbf{h} = \begin{bmatrix} h_{AA} & h_{AB} \\ h_{BA} & h_{BB} \end{bmatrix}.$$

Which conditions must be imposed on the four matrix elements if $\mathbf{h}$ is to be Hermitian?

132. The matrix

$$\begin{bmatrix} 1 & S \\ S & 1 \end{bmatrix}$$

represents another operator in the same basis. Which one?

133. To which forms do these two matrices reduce under the three assumptions above?

Adopting assumptions 1–3 we can write equations (54) and (56) as

$$(\alpha - \epsilon)c_A + \beta c_B = 0;$$

$$\beta c_A + (\alpha - \epsilon)c_B = 0. \qquad \text{(IV.57)}$$

This is a *homogeneous* set of linear equations. As such, it has nontrivial solutions (i.e., not identically zero) only if the determinant of the coefficients vanishes:

$$\begin{vmatrix} \alpha - \epsilon & \beta \\ \beta & \alpha - \epsilon \end{vmatrix} = 0. \qquad \text{(IV.58)}$$

134. Try to solve (57) directly: get an expression for c_B for one of the equations and substitute it in the other. What happens?

135. In what sense is a homogeneous set of equations special?

In (57) we know (i.e., we assume that we know) the parameters α and β. We do *not* know ϵ, however, until we have used the condition (58). That is apparently a second-degree equation in ϵ with the solutions

$$\epsilon_1 = \alpha + \beta; \quad \epsilon_2 = \alpha - \beta. \quad\quad \text{(IV.59)}$$

The parameters α and β both turn out to be negative.

136. Substitute $\alpha + \beta$ for ϵ in (57). Does that give a solution for the unknowns c_A and c_B?

137. Substitute $\alpha - \beta$ for ϵ in (57). Does that give a solution for the unknowns c_A and c_B?

What we have achieved so far is to find two possible one-electron energies expressed in the parameters α and β. It remains to find the corresponding coefficients. Since (57) is a homogeneous set of equations, we can only get ratios of these coefficients.

EXERCISES

138. Use the results of Exercise 134 to calculate the ratios c_{A1}/c_{B1} corresponding to ϵ_1 and c_{A2}/c_{B2} corresponding to ϵ_2.

139. Write up the corresponding molecular orbitals ψ_1 and ψ_2. Compare (52). Compare also with the bonding and antibonding molecular orbitals for the hydrogen molecule.

140. Calculate the overlap integral of ψ_1 and ψ_2, using assumption 1 above.

141. Normalize ψ_1 and ψ_2, using assumption 1 above.

The results of the latest exercises can be expressed in matrix form:

$$[\psi_1 \quad \psi_2] = [\phi_A \quad \phi_B] \begin{bmatrix} c_{A1} & c_{A2} \\ c_{B1} & c_{B2} \end{bmatrix}. \qquad (IV.60a)$$

With obvious notations for the three matrices, this can also be written

$$\psi = \Phi c. \qquad (IV.60b)$$

142. Write up the matrix $\mathbf{c}$ explicitly and show that $\mathbf{c}^{+}\mathbf{c} = \mathbf{c}\mathbf{c}^{+} = \mathbf{1}$. What is a matrix with that property called?

143. Use the result of Exercise 142 to express the atomic orbitals in terms of the molecular orbitals.

After a calculation of this kind it is very important to go back and interpret the results.

144. What is the total number of electrons in the ethene molecule?

145. How many of these are core electrons?

146. How many are used in the σ-skeleton?

147. How many are π-electrons?

For the π-electrons we thus have two MOs available. Each orbital can be occupied by two electrons, one of each spin. The criterion for filling up orbitals is always minimization of the total energy. But here we have a problem. We work entirely in a one-electron approximation and the vital connection to the many-electron Hamiltonian is missing. In this predicament one just adds up the one-electron energies of those orbitals that are occupied in order to get an approximate total energy. That is one of the more debatable procedures in the Hückel method.

With such a criterion we apparently get the lowest possible total energy by filling the available orbitals from the bottom. For the π-electrons in ethene this means that we minimize the total energy by letting both electrons occupy ψ_1.

EXERCISES

148. Write up explicitly those *spin orbitals* that are occupied by π-electrons.

149. What then, is the total energy?

So far we have studied only the ground state. The Hückel method also provides—within its assumptions—a possibility to describe excited states. With four spin orbitals (which ones?) and two electrons, we can construct six possible states (why six?). Some of those will not correspond to antisymmetric two-electron functions, however, and must therefore be excluded.

EXERCISES

150. Calculate the total π-electron energy for the following occupations: **(a)** $\psi_1\alpha\psi_2\alpha$; **(b)** $\psi_1\alpha\psi_2\beta$; **(c)** $\psi_2\alpha\psi_2\beta$.

151. Which are the corresponding excitation energies (i.e., the differences between the total energy of the excited states in Exercise 150 and that of the ground state)?

Butadiene

We continue with a molecule containing four π-electrons. In the formula only the carbon atoms and the bonds between them are shown. Since carbon is four-valued, one can easily find out how many hydrogen atoms there are on each carbon:

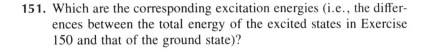

In the figure we mark the conventional double and single bonds. As we will see from the quantum mechanical treatment, the bonding situation is a little more complicated.

EXERCISES

152. How many and which nuclei make up the butadiene molecule?

153. How many core electrons does this molecule contain?

154. How many electrons are there in the σ-skeleton?

155. What is the total number of electrons in the butadiene molecule?

We proceed as in the ethene case from an assumed effective one-electron equation for the π-electrons [cf. (51)]:

$$\mathbf{h}\psi = \epsilon\psi. \tag{IV.61}$$

Now we have four AOs in the basis, ϕ_A, ϕ_B, ϕ_C, ϕ_D, with the subscripts referring to the four carbon atoms.

EXERCISES

156. Write up explicit expressions for these four orbitals.

157. Make a sketch of the σ-skeleton and account for the 18 σ-electrons.

From the four AOs we can form four MOs of the type

$$\psi = \phi_A c_A + \phi_B c_B + \phi_C c_C + \phi_D c_D. \tag{IV.62}$$

Since we are going to neglect all overlap integrals between the AOs (cf. assumption 1), (62) means that we are looking for a unitary transformation $\mathbf{C}$ of the orthonormal set

$$\mathbf{\Phi} = [\phi_A \quad \phi_B \quad \phi_C \quad \phi_D]. \tag{IV.63}$$

The coefficients in (62) will be determined so as to satisfy (61) within the chosen basis.

158. What does the term *unitary* mean?

159. What characterizes a unitary matrix $\mathbf{C}$ of order 4?

Substituting (62) in (61) and proceeding as in the ethene case (verify all the steps carefully), we get after some manipulation:

$$(\alpha - \epsilon)c_A + \beta c_B = 0;$$

$$\beta c_A + (\alpha - \epsilon)c_B + \beta c_C = 0;$$

$$\beta c_B + (\alpha - \epsilon)c_C + \beta c_D = 0;$$

$$\beta c_C + (\alpha - \epsilon)c_D = 0. \qquad \text{(IV.64)}$$

Notice that the structure of these equations reflects the geometrical (or rather, the topological) structure of the molecule. Nontrivial solutions (i.e., not all $c_i = 0$) require the vanishing of the secular determinant:

$$\begin{vmatrix} \alpha - \epsilon & \beta & 0 & 0 \\ \beta & \alpha - \epsilon & \beta & 0 \\ 0 & \beta & \alpha - \epsilon & \beta \\ 0 & 0 & \beta & \alpha - \epsilon \end{vmatrix} = 0. \qquad \text{(IV.65)}$$

For larger systems there are computer programs to find the eigenvalues—the values of ϵ that satisfy (65). In the present case we can still do it by hand. We expand the determinant according to the first row:

$$(\alpha - \epsilon) \begin{vmatrix} \alpha - \epsilon & \beta & 0 \\ \beta & \alpha - \epsilon & \beta \\ 0 & \beta & \alpha - \epsilon \end{vmatrix} - \beta \begin{vmatrix} \beta & \beta & 0 \\ 0 & \alpha - \epsilon & \beta \\ 0 & \beta & \alpha - \epsilon \end{vmatrix}$$

$$= (\alpha - \epsilon)[(\alpha - \epsilon)^3 - 2\beta^2(\alpha - \epsilon)]$$

$$- \beta[\beta(\alpha - \epsilon)^2 - \beta^3]$$

$$= (\alpha - \epsilon)^4 - 3\beta^2(\alpha - \epsilon)^2 + \beta^4 = 0. \qquad (IV.66)$$

The solutions of (66) are

$$\epsilon_1 = \alpha + \frac{\beta(1 + \sqrt{5})}{2} = \alpha + 1.618\beta;$$

$$\epsilon_2 = \alpha + \frac{\beta(1 - \sqrt{5})}{2} = \alpha + 0.618\beta;$$

$$\epsilon_3 = \alpha - \frac{\beta(1 - \sqrt{5})}{2} = \alpha - 0.618\beta;$$

$$\epsilon_4 = \alpha - \frac{\beta(1 + \sqrt{5})}{2} = \alpha - 0.618\beta. \qquad (IV.67)$$

EXERCISES

160. Which occupation of these MOs gives the lowest total energy? Calculate that energy.

161. The result of the Hückel treatment of ethene can be interpreted by saying that the energy of a π-bond with two electrons is $2(\alpha + \beta)$. To which energy, then, does the traditional picture (see the figure) of butadiene correspond?

162. How much energy is thus gained by allowing the four π-electrons to be delocalized throughout the molecule? In what way is the sign of the parameter β important for that result?

Having obtained the four roots (67) of the secular equation (65), we can go back to the linear set of equations (64) and calculate the ratios of the coefficients for each root ϵ_k. The results can be summarized in the matrix

$$\mathbf{C} = \begin{bmatrix} 0.372 & -0.601 & -0.601 & -0.372 \\ 0.601 & -0.372 & 0.372 & 0.601 \\ 0.601 & 0.372 & 0.372 & -0.601 \\ 0.372 & 0.601 & -0.601 & 0.372 \end{bmatrix}, \quad \text{(IV.68)}$$

which provides the transformation from AOs to MOs according to the formula [cf. (63)]

$$\boldsymbol{\psi} = \boldsymbol{\Phi}\mathbf{C}. \quad \text{(IV.69)}$$

The row matrix

$$\boldsymbol{\psi} = [\psi_1 \quad \psi_2 \quad \psi_3 \quad \psi_4] \quad \text{(IV.70)}$$

thus contains the four possible MOs.

EXERCISES

163. Use the previous expressions to write up the four MOs explicitly in terms of the AOs.

164. Calculate the overlap integrals

$$\int dv\ \psi_1^*\psi_2 \quad \text{and} \quad \int dv\ \psi_1^*\psi_4.$$

165. Are the four MOs ψ_k normalized?

Cyclobutadiene

It is instructive to carry through a similar analysis for the cyclic isomer

where we assume that the four C—C distances are equal.

EXERCISES

166. Set up the secular determinant corresponding to (65) for cyclo-butadiene. Show that this determinant is *cyclic* in the sense that the second row can be obtained from the first by cyclic permutations of the elements, the third row from the second in the same way, and so on.

167. Calculate the roots of the secular equation and show that they are

$$\epsilon_1 = \alpha + 2\beta; \quad \epsilon_2 = \epsilon_3 = \alpha; \quad \epsilon_4 = \alpha - 2\beta.$$

168. Calculate the matrix corresponding to (68) for cyclobutadiene.

Benzene

The formula for C_6H_6 with a ring in the middle

illustrates very well the chemical bonding in that molecule. The σ-skeleton (of which only the $C-C$ bonds are shown) is represented by the hexagon and the delocalized π-bonds by the ring. We are now going to see how even the simple Hückel model accounts quite well for this situation.

EXERCISES

169. What is the total number of electrons in the benzene molecule?

170. How many core electrons are there in the benzene molecule?

171. How many in the σ-skeleton are there in the benzene molecule?

172. How many π-electrons are there in the benzene molecule?

As usual in the Hückel model, we assume the existence of an effective one-electron Hamiltonian and we construct approximate eigenfunctions (MOs) of that Hamiltonian [cf. (51)] as LCAOs:

$$\psi = \sum_{\mu=1}^{6} \phi_\mu c_\mu. \tag{IV.71}$$

Substituting (71) in (51) and proceeding as before, we end up with a set of six coupled linear equations,

$$\sum_{\nu=1}^{6} [h_{\mu\nu} - \epsilon\Delta_{\mu\nu}]c_\nu = 0; \qquad \mu = 1, 2, \ldots, 6. \tag{IV.72}$$

EXERCISES

173. Simplify (72) by means of the three "Hückel assumptions."

174. Write up the resulting secular equation, that is, the determinantal condition for nontrivial solutions of (72). Notice the similarity with the corresponding determinant for cyclobutadiene (Exercise 166).

175. How is the cyclic structure of the benzene molecule reflected in the structure of the secular determinant?

Even though it is certainly possible to expand the secular determinant for benzene explicitly as we have done in the other cases, there are much more efficient methods, in particular for a system with high symmetry such as C_6H_6. We consider the following linear combinations of the AOs:

$$\chi_\kappa(\mathbf{r}) = \frac{1}{\sqrt{6}} \sum_{\mu=1}^{6} \phi_\mu(\mathbf{r}) e^{2\pi i \kappa \mu/6}; \qquad \kappa = 1, 2, \ldots, 6. \quad \text{(IV.73)}$$

176. Write up the six functions (73) explicitly; use, for example, the notation $\omega = \exp(i\pi/3) = 1/2 + i\sqrt{3}/2$.

177. Is it meaningful to let κ in (73) take higher values than six or lower values than 1?

178. Normalize the functions (73) assuming that the AOs $\phi_\mu(\mathbf{r})$ are orthonormal. Recall what the subscript μ stands for.

179. Calculate the overlap integral of $\chi_{\kappa'}(\mathbf{r})$ and $\chi_\kappa(\mathbf{r})$ for $\kappa' \neq \kappa$. Note that in general the functions (73) are complex.

The functions (73) have interesting symmetry properties with respect to rotations, which leave the hexagonal skeleton in benzene unchanged. We introduce an operator $\mathbf{T}$ that transforms the AO $\phi_\mu(\mathbf{r})$ to $\phi_{\mu-1}(\mathbf{r})$: $\mathbf{T}\phi_\mu(\mathbf{r}) = \phi_{\mu-1}(\mathbf{r})$.

EXERCISES

180. Which operator is $\mathbf{T}^6$?

181. What does the operator $\mathbf{T}^{-1}$ do?

182. What is $\mathbf{T}\phi_1(\mathbf{r})$?

We apply the operator $\mathbf{T}$ to the functions (73):

$$
\begin{aligned}
\mathbf{T}\chi_\kappa(\mathbf{r}) &= \frac{1}{\sqrt{6}} \sum_{\mu=1}^{6} \mathbf{T}\phi_\mu(\mathbf{r})\, e^{2\pi i\kappa\mu/6} \\
&= \frac{1}{\sqrt{6}} \sum_{\mu'=1}^{6} \phi_{\mu'}(\mathbf{r})\, e^{2\pi i\kappa(\mu'+1)/6} = \chi_\kappa(\mathbf{r})\, e^{2\pi i\kappa/6}. \quad \text{(IV.74)}
\end{aligned}
$$

EXERCISES

183. The second line of (74) has been obtained by changing the summation index from μ to $\mu' = \mu + 1$. How, then, is it possible that μ' takes the same values in the sum as μ?

184. Explain the last step in (74).

185. Use (74) to calculate $\mathbf{T}^2\chi_\kappa$, $\mathbf{T}^3\chi_\kappa$, $\mathbf{T}^4\chi_\kappa$, $\mathbf{T}^5\chi_\kappa$, and $\mathbf{T}^6\chi_\kappa$.

Thus the functions (73) are eigenfunctions of the operator $\mathbf{T}$ with eigenvalue $e^{i\pi\kappa/3} = \omega^\kappa$. Instead of expanding the MOs in the AOs ϕ_μ as in (71), we can expand them in symmetry-adapted functions (73):

$$\psi(\mathbf{r}) = \sum_{\kappa=1}^{6} \chi_\kappa(\mathbf{r})d_\kappa. \tag{IV.75}$$

We then get another set of linear equations for the coefficients d_κ, with $h_{\mu\nu}$ and $\Delta_{\mu\nu}$ in (72) replaced by $h_{\kappa\kappa'}$ and $\Delta_{\kappa\kappa'}$.

EXERCISES

186. Express the overlap integrals $\Delta_{\kappa\kappa'}$ in terms of the original overlap integrals $\Delta_{\mu\nu}$.

187. Calculate $h_{\kappa\kappa'} = \int dv\, \chi_{\kappa'}^* \mathbf{h}\chi_\kappa$ using the ordinary Hückel assumptions.

188. What do the results of Exercises 186 and 187 imply for the equations to be satisfied by the coefficients d_κ in (75)? Write up these equations explicitly for $\kappa = 1, 2, 3, 4, 5$, and 6.

In other words, the introduction of the symmetry-adapted orbitals (74) has solved the problem completely. This is more than we can normally hope for. Symmetry simplifies most problems but usually not "completely."

EXERCISES

189. The transformation (73) can be written in terms of matrices. Write the two sets of functions as row matrices:

$$\chi = [\chi_1, \chi_2, \chi_3, \chi_4, \chi_5, \chi_6]; \qquad \phi = [\phi_1, \phi_2, \phi_3, \phi_4, \phi_5, \phi_6].$$

Express the transformation as a 6×6 matrix $T: \chi = \phi T$. Write up the matrix T explicitly.

190. Show that $T^+T = TT^+ = 1$ (i.e., that the transformation matrix in Exercise 189 T is unitary). That property of the transformation matrix is associated with certain properties of the two sets of functions. Which ones?

Armed with the eigenvalues obtained in Exercise 187, we can calculate the total π-electron energy of benzene in the Hückel approximation:

$$E_{Tot} = 2(\alpha + 2\beta) + 4(\alpha + \beta) = 6\alpha + 8\beta. \quad \text{(IV.76)}$$

EXERCISES

191. How is the total energy in (76) obtained?

192. The ethene example shows that in the Hückel model one π-bond is associated with the energy $\alpha + \beta$. Is the total energy (76) higher or lower than the energy of six π-bonds? Is anything gained by the ring delocalization?

It is also important to use the information inherent in the wave function. In a one-electron method like the Hückel model, such information is available in the MOs, in particular those that are occupied.

EXERCISES

193. Write up explicitly those spin orbitals that are occupied in the ground state of benzene. Which one-electron energy is associated with each of them?

194. Is that information used in the calculation of the total energy above?

195. Form two new linear combinations of the two MOs χ_1 and χ_5 which correspond to the degenerate energy level:

$$\chi_1' = \chi_1 \cos \gamma + \chi_5 \sin \gamma; \qquad \chi_5' = \chi_1 \sin \gamma - \chi_5 \cos \gamma.$$

Assume that χ_1 and χ_5 are normalized and normalize the new functions χ_1' and χ_5'. Calculate their overlap integral. Calculate the Hückel energies of the new orbitals. Comments?

The two molecular situations treated in this chapter although typical of important classes of molecules, obviously cover only a small part of the immensely rich field of molecules. The descriptions given should provide a reasonable starting point for further work, though.

REFERENCES

Atkins, P. W., *Molecular Quantum Mechanics*, Oxford University Press, Oxford, 1970; second edition, 1983.

Cooper, D. L., J. Gerratt, and M. Raimondi, *Chem. Rev.* **91**, 929 (1991).

Slater, J. C., *Quantum Theory of Molecules and Solids*, Vol. I, McGraw-Hill, New York, 1963.

V

POLYMERS

V.A. BASIC ASPECTS OF TRANSLATION SYMMETRY

In the present chapter we are going to study a type of system that in a certain sense is intermediate between small molecules and solids. The structure of polymers can be characterized at several different levels. From the point of view of the electronic structure a polymer can be regarded as a one-dimensional extended system. A polymer is obviously a three-dimensional object. But since its primary structure consists of a very large number of basic units—the monomers—the polymer can be described as a *lattice in one dimension*. This fact has important consequences for the boundary conditions that should be imposed on all wave functions for polymers.

In all directions perpendicular to the direction of the lattice, a polymer "behaves" like a small molecule: when we depart from the "axis" of the polymer the wave functions must go to zero as they do far away from an atom or a small molecule. *Along the axis* the polymer is, however, an extended system, and then we meet a new type of boundary condition. The term *extended system* just means that throughout the system the wave functions for the electrons should be *nonvanishing*. There are basically two ways of satisfying that requirement. We could treat an infinite system directly—which is possible but not very convenient because of the convergence aspects, which must be dealt with very carefully. The standard procedure is, instead, to impose so-called *periodic* or *Born–von Kármán boundary conditions*. One works with a finite but very large piece of the polymer. "Large" means large relative to the size of the monomer. For this finite number of monomers one derives all relations that are needed. In principle, one can then go to the so-called thermodynamic limit. That means basically that both the number of electrons and the volume of the system tend to infinity while the electronic density is kept finite.

This limiting procedure must be carried out with great care primarily because of the long-range nature of the Coulomb forces. What is usually done is first to introduce a fictitious screened Coulomb interaction exp $(-\lambda r)/r$, characterized by a screening constant λ. This leads to expressions that can be handled mathematically without convergence problems. In the final expressions for the total energy, one then carries out two limiting procedures. The size of the system is allowed to grow to infinity at the same time as the screening constant tends to zero, *in such a way that the interaction gets a chance to remain operative throughout the system.* The theoretical tools developed in the present chapter are essential for such a program, even though the full treatment is beyond the scope of the book.

It is easier to grasp the idea of periodic boundary conditions for a one-dimensional lattice than for the three-dimensional situation in a solid. Apart from being quite interesting objects in their own right, polymers are therefore also ideal for introducing the fundamental concepts and procedures needed in solid-state theory. We introduce a coordinate system with the z-axis as the "polymer axis." The length of the unit cell is denoted by a.

EXERCISES

1. What is the difference between the two concepts "unit cell" and "monomer"?

2. If $\psi(\mathbf{r})$ is the value of a wave function in the cell around the origin (see the figure below), what is its value in those cells that are labeled **(a)** $+1$, **(b)** -1, **(c)** -5; and **(d)** $+8$?

Cell Nr -2 -1 0 $+1$ $+2$

The unit cells can be chosen in many different ways even though in most cases one choice seems more convenient than others. Here we use the convention shown in the figure, which thus means that the border planes that separate different cells are found at $z = \ldots, -2a, -a, 0, +a, +2a, \ldots$. We will work with a piece of polymer of length Na and we label the cells in the periodically repeated Born—von Kármán region (BK) as follows:

$$-\frac{N}{2} \leq \mu < \frac{N}{2}. \tag{V.1}$$

EXERCISES

3. Which values does μ take when **(a)** $N = 4$ and **(b)** $N = 6$?

4. What would it mean if $\mu = N$ or $-N$?

The periodic boundary conditions mean that *all* wave functions to be used for the polymer must satisfy

$$\psi(\mathbf{r} - Na\mathbf{e}_z) = \psi(\mathbf{r}). \tag{V.2}$$

EXERCISES

5. Is there any relation between the values of a function satisfying (2) for $\mathbf{r} = (Na/2)\mathbf{e}_z$ and $\mathbf{r} = -(Na/2)\mathbf{e}_z$?

6. What is the value of the function in (2) for $\mathbf{r} = Na\mathbf{e}_z$? That question can be expressed in another way: which value of $\mathbf{r}$ in the BK region corresponds to $Na\mathbf{e}_z$?

For our one-dimensional lattice the periodic boundary conditions can easily be visualized by considering *cyclic systems*. Instead of working with a linear system, we join the two ends of the chain to a big circle. Then cell $nr(-N/2)$ is also cell $nr(-N/2) + N = N/2$, cell $nr - (N/2) + 1$ is also cell $nr(N/2) + 1$, and so on. In other words, the polymer lattice can be viewed as a generalization of the benzene molecule (cf. the end of Chapter IV): instead of six links we have N.

We need an operator corresponding to the $\mathbf{T}$ used in the treatment of benzene. We define the *translation operator* $\mathbf{T}$ by its effect on an arbitrary function:

$$\mathbf{T}\psi(\mathbf{r}) = \psi(\mathbf{r} - a\mathbf{e}_z). \qquad (\text{V.3})$$

EXERCISES

7. Calculate the following functions in terms of $\psi(\mathbf{r})$: **(a)** $\mathbf{T}^2\psi(\mathbf{r})$, **(b)** $\mathbf{T}^3\psi(\mathbf{r})$, and **(c)** $\mathbf{T}^\mu\psi(\mathbf{r})$.

8. What is $\mathbf{T}^N\psi(\mathbf{r})$ for a function $\psi(\mathbf{r})$ satisfying the periodic boundary condition (2)?

Since $\mathbf{r} - Na\mathbf{e}_z$ in our cyclic system is the same point as $\mathbf{r}$, we have the operator equality

$$\mathbf{T}^N = \mathbf{1} \qquad (\text{V.4})$$

for the translation operators defined by (3) when they work on functions satisfying (2). The equality (4) can be used to derive the eigenvalues of **T**. To do so, we start out from the eigenvalue problem

$$\mathbf{T}\chi(\mathbf{r}) = \lambda\chi(\mathbf{r}),\qquad\text{(V.5)}$$

that is, we introduce the notation $\chi(\mathbf{r})$ for an eigenfunction of the translation operator with eigenvalue λ. Like all wave functions for the polymer the eigenfunctions $\chi(\mathbf{r})$ must satisfy the periodic boundary conditions (2).

EXERCISES

9. Use (5) to express the functions $\mathbf{T}^2\chi(\mathbf{r})$, $\mathbf{T}^3\chi(\mathbf{r})$, . . . , $\mathbf{T}^N\chi(\mathbf{r})$ in terms of $\chi(\mathbf{r})$ and λ.

10. Combine (4) with the result of $\mathbf{T}^N\chi(\mathbf{r})$ to calculate the *number* λ^N.

Operating on (5) repeatedly with the operator **T**, we have

$$\mathbf{T}^2\chi(\mathbf{r}) = \lambda\mathbf{T}\chi(\mathbf{r}) = \lambda^2\chi(\mathbf{r});\qquad\text{(V.6}a\text{)}$$

$$\mathbf{T}^3\chi(\mathbf{r}) = \lambda^2\mathbf{T}\chi(\mathbf{r}) = \lambda^3\chi(\mathbf{r});\qquad\text{(V.6}b\text{)}$$

. . . .

$$\chi(\mathbf{r}) = \mathbf{1}\chi(\mathbf{r}) = \mathbf{T}^N\chi(\mathbf{r}) = \lambda^N\chi(\mathbf{r}).\qquad\text{(V.6}c\text{)}$$

Thus the operator relation (4) implies for the eigenvalue λ that unless $\chi(\mathbf{r}) = 0$,

$$\lambda^N = 1.\qquad\text{(V.7)}$$

This apparently shows that the eigenvalues of the translation operator are the N roots of unity, which we label by an integer κ:

$$\lambda_\kappa = e^{-2\pi i \kappa/N}; \qquad -\frac{N}{2} \leq \kappa < \frac{N}{2}. \qquad \text{(V.8)}$$

EXERCISES

11. Calcualte λ_κ for $N = 4$, 6, and 10. Mark these eigenvalues on the unit circle in the complex plane. Are there any eigenvalues that appear for all three values of N?

12. Which are the eigenvalues of the operator $\mathbf{T}^2$?

13. What does the eigenvalue relation (5) actually mean? Spell it out explicitly. Is there any eigenvalue for which the corresponding eigenfunction has the same value in all unit cells?

14. The interval chosen for the integer κ in (8) is arbitrary. Instead, label with a subscript κ' defined for $0 \leq \kappa' < N - 1$. Find the relations between κ and κ'.

The eigenvalues (8) and therefore also the eigenfunctions are thus labeled by an integer κ, which can take as many different values as the number of cells in the lattice:

$$\mathbf{T}\chi_\kappa(\mathbf{r}) = \chi_\kappa(\mathbf{r} - a\mathbf{e}_z) = \lambda_\kappa\chi_\kappa(\mathbf{r}) = e^{-2\pi i\kappa/N}\chi_\kappa(\mathbf{r}). \quad (V.9)$$

This means that when we go from a point in one unit cell to the corresponding point in another cell, the eigenfunction is multiplied by a complex number of length 1—a phase factor. It is very important to notice that in general these eigenfunctions are *not periodic*, when we go from one cell to another one.

EXERCISES

15. Do the functions $\chi_\kappa(\mathbf{r})$ in (9) satisfy the periodic boundary condition (2)?

16. Which relation is satisfied by a function $f(\mathbf{r})$ with the periodicity of the lattice?

V.B. CONSTRUCTION OF EIGENFUNCTIONS OF THE TRANSLATION OPERATOR

The eigen*values* of the translation operator are thus known as soon as we have chosen the size Na of the BK region. To determine the corresponding eigenfunctions is a little more involved but not particularly difficult. We demonstrate here how one can construct such functions from arbitrary functions by means of so-called *projection operators*.

One road to these projection operators starts out from the eigenvalue relation (9), which we rewrite slightly as

$$[\mathbf{T} - \lambda_\kappa \mathbf{1}]\chi_\kappa(\mathbf{r}) = 0. \quad (V.10)$$

A new interpretation of (10) is to say that the *annihilator* (or annihilation operator) $\mathbf{T} - \lambda_\kappa\mathbf{1}$ annihilates the function $\chi_\kappa(\mathbf{r})$.

---

17. What happens when $\mathbf{T} - \lambda_\kappa \mathbf{1}$ operates on the function $\chi_{\kappa+1}(\mathbf{r})$?

18. Do the two annihilators $[\mathbf{T} - \lambda_\kappa \mathbf{1}]$ and $[\mathbf{T} - \lambda_{\kappa+1}\mathbf{1}]$ commute?

19. What happens when the product $[\mathbf{T} - \lambda_\kappa \mathbf{1}][\mathbf{T} - \lambda_{\kappa+1}\mathbf{1}]$ operates on the function $\chi_{\kappa+1}(\mathbf{r})$?

An arbitrary (in the sense that it does not have any particular symmetry properties with respect to $\mathbf{T}$) function $\phi(\mathbf{r})$ can be thought of as a mixture of (in principle) all possible eigenfunctions of $\mathbf{T}$ with unknown coefficients:

$$\phi(\mathbf{r}) = \sum_{\kappa=-N/2}^{N/2-1} \chi_\kappa(\mathbf{r}) c_\kappa. \qquad (V.11)$$

---

20. Which properties do the two possible eigenfunctions of $\mathbf{T}$ have in the case $N = 2$?

21. An arbitrary function $\phi(\mathbf{r})$ can be written as a linear combination of the two functions

$$\phi_+(\mathbf{r}) = \tfrac{1}{2}[\phi(\mathbf{r}) + \phi(\mathbf{r} - \mathbf{a})];$$

$$\phi_-(\mathbf{r}) = \tfrac{1}{2}[\phi(\mathbf{r}) - \phi(\mathbf{r} - \mathbf{a})].$$

Find the explicit coefficients of these two functions in $\phi(\mathbf{r})$. How do $\phi_+(\mathbf{r})$ and $\phi_-(\mathbf{r})$ transform under the operation of the translation operator?

22. What happens (still when $N = 2$) when the annihilators $[\mathbf{T} - \lambda_{-1}\mathbf{1}]$ and $[\mathbf{T} - \lambda_0\mathbf{1}]$ operate on $\phi(\mathbf{r})$?

Each annihilator kills a particular term in a sum like (11). If we are interested in the eigenfunction with a particular label $\kappa = p$, we should get rid of all the other components:

$$[\mathbf{T} - \lambda_{-N/2}\mathbf{1}][\mathbf{T} - \lambda_{-N/2+1}\mathbf{1}]$$

$$\cdots [\mathbf{T} - \lambda_{p-1}\mathbf{1}][\mathbf{T} - \lambda_{p+1}\mathbf{1}]$$

$$\cdots [\mathbf{T} - \lambda_{N/2-1}\mathbf{1}]\phi(\mathbf{r})$$

$$= \chi_p(\mathbf{r})c_p[\lambda_p - \lambda_{-N/2}][\lambda_p - \lambda_{-N/2+1}]$$

$$\cdots [\lambda_p - \lambda_{N/2-1}]. \qquad (\text{V}.12)$$

Since we are not interested in the product of eigenvalue differences, we define an operator

$$\mathbf{0}_p = \prod_{j \neq p} \frac{\mathbf{T} - \lambda_j\mathbf{1}}{\lambda_p - \lambda_j}, \qquad (\text{V}.13)$$

which apparently gives

$$\mathbf{0}_p\phi(\mathbf{r}) = \chi_p(\mathbf{r})c_p. \qquad (\text{V}.14)$$

In other words, the function $\mathbf{0}_p\phi(\mathbf{r})$ is an eigenfunction of $\mathbf{T}$ with eigenvalue λ_p. It is very important that this statement be true for an *arbitrary* function $\phi(\mathbf{r})$, unless the coefficient c_p in the expansion (11) vanishes. In that case the trial function $\phi(\mathbf{r})$ simply does not contain the component $\chi_p(\mathbf{r})$.

The operator $\mathbf{0}_p$ is called a *projection operator* since it gives the same result if it operates twice or more times (how is that possible)? To understand the reason for this term, it is instructive to carry out a projection geometrically in two or three dimensions, *repeatedly*.

EXERCISES

23. What happens when the product of *all* the N possible annihilators works on an arbitrary function?

24. The answer to Exercise 23 can be expressed as an operator identity for the operator

$$\prod_{j=-N/2}^{N/2-1} [\mathbf{T} - \lambda_j \mathbf{1}]. \qquad \text{Which one?}$$

The important relation

$$\mathbf{0}_p^2 = \mathbf{0}_p, \qquad (V.15)$$

can be demonstrated by writing a typical factor in (13) as

$$\frac{\mathbf{T} - \lambda_j \mathbf{1}}{\lambda_p - \lambda_j} = \mathbf{1} + \frac{\mathbf{T} - \lambda_p \mathbf{1}}{\lambda_p - \lambda_j}. \qquad (V.16)$$

Then we get (verify each step carefully)

$$\mathbf{0}_p^2 = \mathbf{0}_p \prod_{j \neq p} \left[\mathbf{1} + \frac{\mathbf{T} - \lambda_p \mathbf{1}}{\lambda_p - \lambda_j} \right] = \mathbf{0}_p + 0 = \mathbf{0}_p. \qquad (V.17)$$

The "0" in (17) should more properly be written as the number

0 times the identity operator: $0 \cdot \mathbf{1}$. We can also use the result of the Exercises 23 and 24 to show that for $p \neq q$,

$$\mathbf{0}_p \mathbf{0}_q = 0. \qquad \text{(V.18)}$$

EXERCISES

25. Illustrate this with the case $N = 4$. **(a)** Which are the four eigenvalues? **(b)** Write up the four projection operators and show explicitly that $\mathbf{0}_{-1}^2 = \mathbf{0}_{-1}$; $\mathbf{0}_0^2 = \mathbf{0}_0$.

26. What are **(a)** $\mathbf{0}_{-2}\mathbf{0}_{-1}$; **(b)** $\mathbf{0}_{-2}\mathbf{0}_0$; and **(c)** $\mathbf{0}_0\mathbf{0}_1$?

27. What is $\mathbf{0}_{-2} + \mathbf{0}_{-1} + \mathbf{0}_0 + \mathbf{0}_1$?

The results of Exercises 25–27 can be used to write the four projection operators for $N = 4$ in a different form as

$$\mathbf{0}_{-2} = \tfrac{1}{4}[\mathbf{T}^{-2} - \mathbf{T}^{-1} + \mathbf{1} + \mathbf{T}]; \qquad \text{(V.19}a)$$

$$\mathbf{0}_{-1} = \tfrac{1}{4}[-\mathbf{T}^{-2} - i\mathbf{T}^{-1} + \mathbf{1} + i\mathbf{T}]; \qquad \text{(V.19}b)$$

$$\mathbf{0}_0 = \tfrac{1}{4}[\mathbf{T}^{-2} + \mathbf{T}^{-1} + \mathbf{1} + \mathbf{T}]; \qquad \text{(V.19}c)$$

$$\mathbf{0}_1 = \tfrac{1}{4}[-\mathbf{T}^{-2} + i\mathbf{T}^{-1} + \mathbf{1} - i\mathbf{T}]. \qquad \text{(V.19}d)$$

What we see here is another form of the projection operator (13), namely,

$$\mathbf{0}_p = \frac{1}{N} \sum_{\mu=-N/2}^{N/2-1} \lambda_p^{-\mu} \mathbf{T}^\mu. \qquad \text{(V.20)}$$

139

28. Show that if both κ and κ' lie between $-N/2$ and $N/2$,

$$\frac{1}{N} \sum_{\nu=-N/2}^{N/2-1} \exp\left[\frac{2\pi i \nu (\kappa - \kappa')}{N}\right] = \delta_{\kappa\kappa'}.$$

29. What is the value of the previous sum for arbitrary integers κ and κ'?

30. Show, using the form (20) of the projection operators, that

$$\mathbf{0}_p \mathbf{0}_q = \delta_{pq} \mathbf{0}_p.$$

We use the form (20) to show the important *resolution of the identity*

$$\sum_{p=-N/2}^{N/2-1} \mathbf{0}_p = \frac{1}{N} \sum_{\mu=-N/2}^{N/2-1} \mathbf{T}^\mu \sum_{p=-N/2}^{N/2-1} e^{-2\pi i p \mu / N}$$

$$= \sum_{\mu=-N/2}^{N/2-1} \mathbf{T}^\mu \delta_{\mu 0} = \mathbf{1}. \qquad (V.21)$$

The reason for the name *resolution of the identity* is apparently that the identity operator can be written as a sum of projection operators, each corresponding to one symmetry type.

31. What is the connection between (21) and (11)?

32. Since any function $\phi(\mathbf{r})$ can be thought of as the identity operator working on it: $\mathbf{1} \cdot \phi(\mathbf{r})$, use (21) to write $\phi(\mathbf{r})$ as a sum of the N possible components. Use that expression to calculate $\mathbf{0}_q \phi(\mathbf{r})$.

33. "Rethink" all the expressions derived in the present section and notice how much one can learn from formal manipulations.

We thus have two forms of the projection operator: (13) and (20). Of these, (20) is probably most convenient for actual applications. In the next chapter we apply it to several types of functions.

V.C. MO-LCAO FUNCTIONS

For polymers the MO-LCAO model is the most important. To get thoroughly familiar with it, we first consider the very simplest form of a polymer, a linear monatomic chain. We choose the unit cells in such a way that the atoms lie in the middle of each cell.

34. What are the coordinates of the endpoints of the cells?

35. What are the coordinates of the atomic position?

36. Center an AO, $\phi(\mathbf{r})$, "on" each atom. Specify the expressions for the N atomic orbitals in the chain.

37. Where is $\phi_0(\mathbf{r})$ centered?

The N AOs, for which we choose the notation $\phi_\mu(\mathbf{r}) = \phi(\mathbf{r} - (\mu + \frac{1}{2})a\mathbf{e}_z)$, can be transformed to N Bloch functions that are adapted to the translational symmetry. For that particular symmetry it is particularly simple to specify what that expression means. A Bloch function is nothing but an eigenfunction of the translation operator. The term *Bloch function* is synonymous with the expression "symmetry adapted to the translational symmetry." As we have seen earlier in this chapter [cf. (8)] the eigenfunctions and eigenvalues of the translation operators are labeled by the N integers κ:

$$\mathbf{T}\chi_\kappa(\mathbf{r}) = e^{-2\pi i\kappa/N}\chi_\kappa(\mathbf{r}); \qquad -\frac{N}{2} \leq \kappa < \frac{N}{2}. \qquad \text{(V.22)}$$

EXERCISES

38. Why does it make sense to say that a Bloch function is an eigenfunction of the translation operators (in the plural)? Which are these operators?

39. Does a Bloch function satisfy the fundamental periodic boundary condition?

40. The "quantum number" κ labels the N possible symmetry types with respect to translations. For two values of κ, namely $\kappa = 0$ and $\kappa = -N/2$, one can easily describe the symmetry of the corresponding Bloch function in words. Do that.

We apply the sum form (20) of the projection operator $\mathbf{0}_\kappa$ to one of the AOs, $\phi_\mu(\mathbf{r}) = \phi(\mathbf{r} - (\mu + \frac{1}{2})a\mathbf{e}_z)$ [e.g., $\phi_0(\mathbf{r})$], which is centered in the middle of cell number 0.

EXERCISES

41. What are $\mathbf{T}\phi_0(\mathbf{r})$, $\mathbf{T}^2\phi_0(\mathbf{r})$, and $\mathbf{T}^5\phi_0(\mathbf{r})$?

42. What is $\mathbf{0}_\kappa\phi_0(\mathbf{r})$?

43. Which function is obtained in Exercise 42 if instead of starting with $\phi_0(\mathbf{r})$, we start with $\phi_4(\mathbf{r})$?

V.D. PROPERTIES OF BLOCH FUNCTIONS

It is worth studying the function

$$\mathbf{0}_\kappa \phi_0(\mathbf{r}) = \frac{1}{N} \sum_{\mu=-N/2}^{N/2-1} e^{2\pi i \kappa \mu/N} \phi_\mu(\mathbf{r}) \qquad (V.23)$$

more closely. First we verify that it is a Bloch function:

$$\mathbf{T0}_\kappa \phi_0(\mathbf{r}) = \frac{1}{N} \sum_{\mu=-N/2}^{N/2-1} e^{2\pi i \kappa \mu/N} \phi_{\mu+1}(\mathbf{r})$$

$$= \frac{1}{N} \sum_{\mu'=-N/2}^{N/2-1} e^{2\pi i \kappa (\mu'-1)/N} \phi_{\mu'}(\mathbf{r}) = e^{-2\pi i \kappa/N} \mathbf{0}_\kappa \phi_0(\mathbf{r}).$$

$$(V.24)$$

EXERCISES

44. Why is it correct to have the same limits for the two summations over μ and μ' in (24)?

45. Write out the two functions $\mathbf{0}_0 \phi_0(\mathbf{r})$ and $\mathbf{0}_{-N/2} \phi_0(\mathbf{r})$ explicitly. What are $\mathbf{T0}_0 \phi_0(\mathbf{r})$ and $\mathbf{T0}_{-N/2} \phi_0(\mathbf{r})$, respectively?

The relation (24) can also be interpreted as a transformation property: under a translation the Bloch function $\mathbf{0}_\kappa \phi_0(\mathbf{r})$ transforms into itself times a complex number depending on the subscript κ. That is a particularly simple form of a transformation property. Usually, one has *sets* of several functions that get transformed into linear combinations of each other under the operation of a symmetry operator.

If a unit vector $\mathbf{e}$ in a plane is projected on the x-axis, the length of the x-component is smaller than or equal to 1. (When is it equal to 1?) A corresponding result holds for functions. The AOs $\phi_\mu(\mathbf{r})$ are assumed to be normalized, and then their projection (23) on the "κ-axis" is in general not normalized. To check that, we first notice that the AOs are in general *not orthogonal*:

$$\int dv \; \phi_\mu^*(\mathbf{r})\phi_\nu(\mathbf{r}) = \Delta_{\mu\nu} \neq 0. \qquad (V.25)$$

EXERCISES

46. Write out the integrand in (25) explicitly when $\phi(\mathbf{r}) = (1/\sqrt{\pi})e^{-r}$ (i.e., the ground-state orbital of the hydrogen atom).

47. What is then $\Delta_{\mu\nu}$? See (IV.8).

48. What is $\Delta_{\mu\mu}$?

To calculate the normalizaiton constant of (23), we can proceed in different ways. The most direct and the most cumbersome, is to use the explicit sum in (23) as it stands. Since there are two factors in the integrand, that means that a double sum will be needed. The property (17) of the projection operator $\mathbf{0}_\kappa$ makes it possible to restrict the operation to a single sum. We also need the turnover rule [cf. (II.37)],

$$\langle \phi | \mathbf{A}\psi \rangle = \langle \mathbf{A}^+\phi | \psi \rangle. \qquad (V.26)$$

EXERCISES

49. Write out what (26) means in terms of explicit integrals.

50. Specialize (26) to the case when **A** is a Hermitian operator.

51. What does (26) mean when **A** is the identity operator?

To calculate the normalization integral of the projected function (23), we proceed as follows:

$$\langle \mathbf{0}_\kappa \phi_0 | \mathbf{0}_\kappa \phi_0 \rangle = \langle \phi_0 | \mathbf{0}_\kappa^+ \mathbf{0}_\kappa \phi_0 \rangle = \langle \phi_0 | \mathbf{0}_\kappa^2 \phi_0 \rangle = \langle \phi_0 | \mathbf{0}_\kappa \phi_0 \rangle.$$

$$(V.27)$$

EXERCISES

52. One of the steps in (27) presupposes that $\mathbf{0}_\kappa$ is a Hermitian operator. Show that despite the fact that the translation operators are *not* Hermitian, $\mathbf{0}_\kappa^+ = \mathbf{0}_\kappa$.

53. Use the sum form (20) of the projection operator to write out its adjoint explicitly. What is the reason for the result $\mathbf{0}_\kappa^+ = \mathbf{0}_\kappa$?

We combine (27) with (23) and (25):

$$\langle \mathbf{0}_\kappa \phi_0 | \mathbf{0}_\kappa \phi_0 \rangle = \frac{1}{N} \sum_{\mu=-N/2}^{N/2-1} e^{2\pi i \kappa \mu / N} \Delta_{0\mu} = N_\kappa. \qquad \text{(V.28)}$$

The normalized Bloch function is thus

$$\psi(\kappa, \mathbf{r}) = \frac{1}{\sqrt{N_\kappa}} \mathbf{0}_\kappa \phi_0(\mathbf{r}). \qquad \text{(V.29)}$$

EXERCISES

54. Calculate N_κ in the case when the AOs are orthonormal.

55. Write the sum in (28) ordered after neighbors in the case when the overlap integrals depend only on the distance between the cells.

56. Calculate (28) in the case when the overlap integrals (25) decay so quickly [cf. (IV.8)] that all except those corresponding to nearest neighbors can be neglected.

One of the most important properties of the symmetry-adapted functions (23) or (29) is that Bloch functions associated with different values of κ are orthogonal. This is what one could expect for eigenfunctions associated with different eigenvalues. Since it is such a fundamental aspect of the Bloch functions, we also study this property from a slightly different point of view.

57. Use the same technique as was used to calculate (28) together with (18) to show that the functions (29) satisfy

$$\int dv \ \psi^*(\kappa', \mathbf{r})\psi(\kappa, \mathbf{r}) = \delta_{\kappa'\kappa}.$$

58. Use the explicit sum (23) to obtain the same result "the long way." Then one sees that a large number of terms cancel. Study in detail how that cancellation occurs.

Most polymers obviously contain more than one atom per unit cell. An interesting "realistic" case, which nevertheless provides important generalizations of the monatomic linear chain, is polyacetylene, $(CH)_x$.

The polyacetylene chain consists of alternating double (short) and single (long) bonds. We introduce the notations d_1 for the short (1.36 Å) and d_2 for the long (1.44 Å) carbon–carbon bond length. The C—H bond length (1.11 Å) is denoted h. We place the borderlines between the unit cells in the middle of the long bond.

59. Calculate the explicit coordinates of the four nuclei in cell number 0 in terms of the bond lengths and the angle α between the long and short bonds.

60. Which relation holds between a, d_1, d_2, and α?

61. What are the coordinates of the two carbon nuclei in cells 1 and -1?

In the π-electron approximation (cf. the treatment of C_2H_4 and C_6H_6 in Chapter IV) we study the electronic structure of the π-electrons in a frozen σ-skeleton.

EXERCISES

62. How do the assumptions about the σ-skeleton check with the geometric structure of polyacetylene?

63. Which orbitals from the free atoms form the σ-bonds in cell 0 and its two neighboring cells?

64. Use formula (IV.8) for the overlap between two $1s$-orbitals centered on two nuclei at a distance R, to calculate the two overlap integrals $\langle H1_0|H2_0\rangle$ and $\langle H2_0|H1_1\rangle$.

Given a cyclic (why?) σ-skeleton with N unit cells the functional raw material thus consists of $2N$ AOs of $2p_z$ type centered at the carbon atoms. We use the notation $\phi(\mathbf{r})$ for such an orbital centered at the origin.

---

65. Write up explicitly the $2p_z$ orbitals that are centered at the two carbon atoms in cell 0.

66. Write up the orbitals in cells 1 and -1.

67. Use the notation $\mathbf{T}$ for the operator which operating on any function $f(\mathbf{r})$ gives $\mathbf{T}f(\mathbf{r}) = f(\mathbf{r} - a\mathbf{e}_z)$. Relate the orbitals used in Exercises 65 and 66 by means of this operator.

Even though the $2p_z$-orbitals on the two carbon atoms in a unit cell are all of the same type, they are nonequivalent in the sense that they are centered at nuclei occupying nonequivalent positions. For these two "types" of AOs we will use the notation $\phi_{1\mu}(\mathbf{r})$ and $\phi_{2\mu}(\mathbf{r})$. We will therefore be able to form two sets of Bloch functions, one from each "type" of atom.

---

68. Use the same technique as for the monatomic chain to form Bloch functions $\mathbf{O}_\kappa \phi_{i0}$, $i = 1, 2$; normalize these functions.

69. Show that $\int dv\ \psi_i^*(\kappa', \mathbf{r})\psi_i(\kappa, \mathbf{r}) = \delta_{\kappa'\kappa}$ for $i = 1, 2$.

70. What is $\int dv\ \psi_1^*(\kappa', r)\psi_2(\kappa, \mathbf{r})$ for **(a)** $\kappa' = \kappa$; **(b)** $\kappa' \neq \kappa$?

Thanks to the symmetry adaptation, we have thus constructed two sets of "internally" orthonormal Bloch functions:

$$\int dv\ \psi_i^*(\kappa', \mathbf{r})\psi_i(\kappa, \mathbf{r}) = \delta_{\kappa'\kappa}. \qquad (V.30)$$

Bloch functions belonging to different sets are orthogonal for different κ, *but not otherwise*.

V.E. MOMENTUM SPACE

It is very instructive to set up the counterparts of the Bloch functions in momentum space, in particular since there is a close connection between momentum space and reciprocal space. The latter concept will be more explicitly defined in Chapter VI, but as we will see, it is just as useful in the polymer case. We first study the situation for the monatomic linear chain. We use (III.5) to introduce the momentum-space counterpart of the basic AO

$$\underline{\phi}(\mathbf{p}) = \frac{1}{\sqrt{8\pi^3}} \int dv\ \phi(\mathbf{r})e^{-i\mathbf{p}\cdot\mathbf{r}}. \qquad (V.31)$$

EXERCISES [Here we use the notation $\phi_\mu(\mathbf{r}) = \phi(\mathbf{r} - (\mu + 1/2)a\mathbf{e}_z)$.]

71. What is the momentum-space counterpart of the functions $\phi_0(\mathbf{r})$, $\phi_1(\mathbf{r})$, $\phi_{-1}(\mathbf{r})$, and $\phi_\mu(\mathbf{r})$?

72. What is the momentum-space counterpart of the functions $\phi_{-N/2}(\mathbf{r})$ and $\phi_{N/2}(\mathbf{r})$?

73. Which relation holds between the two functions in Exercise 72 if periodic boundary conditions have been imposed?

74. What does the result of Exercise 73 imply for the z-component of the momentum?

The periodic boundary conditions (2) that we have imposed on all polymer wave functions in position space lead to a very interesting property for the momentum-space counterparts of those functions. We apply the basic Fourier transform (III.5) to a function $f(\mathbf{r})$ satisfying (2):

$$\underline{f}(\mathbf{p}) = \frac{1}{\sqrt{8\pi^3}} \int dv\, f(\mathbf{r})e^{-i\mathbf{p}\cdot\mathbf{r}}$$

$$= \frac{1}{\sqrt{8\pi^3}} \int dv\, f(\mathbf{r}-Na\mathbf{e}_z)e^{-i\mathbf{p}\cdot\mathbf{r}} = \underline{f}(\mathbf{p})e^{-ip_zNa}.$$

$$(\text{V}.32)$$

That can apparently hold only if either

$$e^{-ip_zNa} = 1, \qquad (\text{V}.33)$$

or $\underline{f}(\mathbf{p}) = 0$. In other words, the momentum-space counterpart $\underline{f}(\mathbf{p})$ of a function $f(\mathbf{r})$ satisfying the periodic boundary conditions (2) *vanishes* unless the z-component of the momentum is of the form

$$p_z = \frac{2\pi\kappa_p}{Na}; \qquad \kappa_p \text{ any positive or negative integer or zero.}$$

$$\text{(V.34)}$$

In this sense the periodic boundary conditions in position space imply a *discretization of momentum space.*

We reinterpret the result of Exercises 71 and 72 as follows:

$$\frac{1}{\sqrt{8\pi^3}} \int dv \; [\mathbf{T}f(\mathbf{r})] e^{-i\mathbf{p}\cdot\mathbf{r}} = \underline{f}(\mathbf{p}) e^{-ip_z a}. \qquad \text{(V.35)}$$

Thus a translation in position space corresponds to multiplication by a phase factor in momentum space. The expression (35) is also related to a more explicit form of the translation operator, which can be obtained from the ordinary Taylor expansion of a function,

$$f(x + h) = f(x) + \frac{h}{1!} f'(x) + \frac{h^2}{2!} f''(x) + \cdots . \qquad \text{(V.36)}$$

We introduce the differentiation operator

$$D = \frac{d}{dx}, \qquad \text{(V.37)}$$

and rewrite (36) as

$$f(x + h) = \left[1 + \frac{hD}{1!} + \frac{(hD)^2}{2!} + \cdots \right] f(x) = e^{hD} f(x).$$

$$\text{(V.38)}$$

In other words, the translation from x to $x + h$ can be achieved by means of the operator e^{hD}.

EXERCISES

75. Use (38) to calculate $f(x + 2h)$. Are the two expressions consistent?

76. Use (38) to calculate $f(x - h)$.

In the polymer case we have used the translation operator (3). Using (38), we thus have

$$\mathbf{T} = \exp{(-aD_z)}, \qquad \text{(V.39)}$$

with

$$D_z = \frac{\partial}{\partial z}. \qquad \text{(V.40)}$$

In the "ordinary" position-space representation the momentum operator is represented by the operator (in atomic units)

$$\mathbf{p} \rightarrow -i\nabla; \quad p_z \rightarrow -i\frac{\partial}{\partial z}. \qquad \text{(V.41)}$$

This implies that the translation operator (39) can be written

$$\mathbf{T} = \exp{(-ip_z a)} = \exp{\left(-a\frac{\partial}{\partial z}\right)}. \qquad \text{(V.42)}$$

The connection between (35) and (42) should be studied carefully. It provides an excellent illustration of two different representations of an operator. In the position representation we need the expression (39) with the differential operator. Since that operator can also be written as (42), we can use it directly as a multiplicative operator in momentum space, which we have already shown explicitly in (35).

EXERCISES

77. What is the counterpart in position space of a function in momentum space which has been translated by a vector $\mathbf{k}$?

78. How does this situation differ from the "inverse" situation we have just studied?

V.F. BLOCH FUNCTIONS IN MOMENTUM SPACE

We apply the result of the preceding section to the basic definition (22) of a Bloch function. Combining (22) and (35), we have

$$\underline{\chi}_\kappa(\mathbf{p})e^{-ip_z a} = \underline{\chi}_\kappa(\mathbf{p})e^{-2\pi i\kappa/N}. \qquad (V.43)$$

This implies that a Bloch function in a momentum space vanishes *unless*

$$\exp\left[i(p_z a - 2\pi\kappa/N)\right] = 1. \qquad (V.44)$$

We know already that *all* functions satisfying periodic boundary conditions in position space vanish unless p_z is of the form (34). For Bloch functions (44) implies the further condition that

$$p_z a - \frac{2\pi\kappa}{N} = \frac{2\pi}{N}(\kappa_p - \kappa) \qquad (V.45)$$

must be an integer times 2π. The difference between the two integers κ_p and κ must therefore be a multiple of N:

$$\kappa_p - \kappa = N\nu; \qquad \nu \text{ positive or negative integer or zero.} \qquad (V.46)$$

That apparently means that the momentum-space Bloch function can be different from zero only for those p_z that satisfy

$$p_z = \frac{2\pi}{Na}[N\nu + \kappa] = \frac{2\pi\nu}{a} + \frac{2\pi\kappa}{Na}. \qquad (V.47)$$

79. Calculate the momentum-space counterpart of the LCAO-type Bloch function (23). Verify that the result satisfies the general rule just obtained.

80. What is the momentum-space counterpart of that Bloch function for $\kappa = 0$?

81. What is the momentum-space counterpart of that Bloch function for $\kappa = -N/2$?

The result of Exercises 79–81 shows that the specific content of the Bloch function is concentrated in the momentum-space counterpart of the AO, which is the "root" of the LCAO Bloch function.

V.G. WANNIER FUNCTIONS

The periodic boundary conditions (2) were introduced to make it possible to construct wave functions that are nonvanishing throughout the Born–von Kármán region. Bloch functions satisfy (2) and it is very instructive to see a little more explicitly why that is the case.

82. How does the defining relation (22) for a Bloch function imply that it is nonvanishing throughout BK?

83. Consider a Bloch function of LCAO type, (29). Let the AO, $\phi(\mathbf{r})$, which is used to form it be of $1s$ type so that $\phi(\mathbf{r})$ vanishes within each unit cell before reaching the cell boundary. Are the statements in the paragraph above compatible with such a situation?

84. Consider a Bloch function of LCAO type constructed from another type of AO, which is different from zero at the cell boundary. Confront such a situation with the earlier statements.

We consider a Bloch function of LCAO type. Such a function can vanish in basically two different situations: (1) when the "mother" AO itself vanishes; (2) when the real or imaginary part of the multiplying factor vanishes. A Bloch function is in general a complex function, even if it is constructed from a purely real AO, and it is therefore natural to study its real and imaginary parts separately.

EXERCISES

85. Write up the real and imaginary parts of a Bloch function such as (29) when the basic AOs are real.

86. If the basic AO has no nodes itself, when do the real and imaginary parts, respectively, of the Bloch function vanish?

87. Plot the real and imaginary parts of these Bloch functions as functions of z for **(a)** $\kappa = 0$; **(b)** $\kappa = -N/2$; **(c)** $\kappa = \pm N/4$.

Thus even if the Bloch functions have characteristic local properties near the nuclei, their global behavior is regulated by the modulating phase coefficients.

The N Bloch functions of a certain type [e.g., $\psi(\kappa, \mathbf{r})$ in (29) for $-N/2 \leq \kappa < N/2$] form an orthonormal set of functions (why?),

$$\int dv \, \psi^*(\kappa', \mathbf{r})\psi(\kappa, \mathbf{r}) = \delta_{\kappa'\kappa}. \qquad \text{(V.48)}$$

From an orthonormal set one can construct other orthonormal sets by means of unitary transformations:

$$\chi(\lambda, \mathbf{r}) = \sum_{\kappa=-N/2}^{N/2-1} \psi(\kappa, \mathbf{r})V_{\kappa\lambda}; \qquad -N/2 \leq \lambda < N/2. \quad \text{(V.49)}$$

EXERCISES

88. Which property must the coefficients $V_{\kappa\lambda}$ in (49) have in order for the new functions $\chi(\lambda, \mathbf{r})$ to form an orthonormal set?

89. What can be said about the matrix product $\mathbf{VV}^+$ if $\mathbf{V}^+\mathbf{V} = \mathbf{1}$?

90. Use Exercise 89 to express the "old" functions ψ in the "new" ones χ.

The results obtained in the previous exercises can be expressed more succinctly if we summarize the N relations (49) in matrix form. The two sets of functions are collected in rows:

$$\mathbf{\Psi} = [\psi(-N/2; \mathbf{r}), \, \psi(-N/2 + 1; \mathbf{r}), \, \dots \, \psi(N/2 - 2; \mathbf{r}),$$

$$\psi(N/2 - 1; \mathbf{r})]; \qquad \text{(V.50a)}$$

$$\chi = [\chi(-N/2; \mathbf{r}), \chi(-N/2 + 1; \mathbf{r}), \ldots \chi(N/2 - 2; \mathbf{r}),$$

$$\chi(N/2 - 1; \mathbf{r})], \tag{V.50b}$$

and the coefficients $V_{\kappa\lambda}$ in a square matrix:

$$\mathbf{V} = \begin{bmatrix} V_{-N/2, -N/2} & V_{-N/2, -N/2+1} & \cdots & V_{-N/2, N/2-1} \\ V_{-N/2+1, -N/2} & V_{-N/2+1, -N/2+1} & \cdots & V_{-N/2+1, N/2-1} \\ \cdots & \cdots & \cdots & \cdots \\ V_{N/2-1, -N/2} & V_{V/2-1, -N/2+1} & \cdots & V_{N/2-1, N/2-1} \end{bmatrix}. \tag{V.51}$$

The N relations (49) can then be written

$$\chi = \mathbf{\Psi V}. \tag{V.52}$$

EXERCISES

91. Set $N = 4$ and write out the matrices corresponding to (50) and (51) explicitly.

92. Verify (in the case $N = 4$) that (52) is equivalent to (49).

93. Multiply (52) by $\mathbf{V}^+$ from the right. What result follows?

The integrations involved in the calculation of overlap integrals can also be expressed in a more condensed form. For that purpose we think of the row matrices (50) as matrices with an infinite set of rows, each labeled by a particular value of the continuous variable **r**, and N columns labeled by the integer λ. The matrix product $\chi^+\chi$ is then a way of denoting the overlap matrix of the set χ, since its elements are the overlap integrals,

$$[\chi^+\chi]_{\lambda'\lambda} = \int dv \, \chi^*(\lambda'; \mathbf{r})\chi(\lambda; \mathbf{r}). \qquad (V.53)$$

Using (52), we then have

$$\chi^+\chi = (\mathbf{\Psi V})^+ \mathbf{\Psi V} = \mathbf{V}^+\mathbf{\Psi}^+\mathbf{\Psi V} = \mathbf{V}^+\mathbf{V} = \mathbf{1}, \quad (V.54)$$

which is one way of stating that the set χ consists of N orthonormal functions.

EXERCISES

94. In (54) we have used the result $\mathbf{\Psi}^+\mathbf{\Psi} = 1$. What is the reason for that expression?

95. Set up explicitly a typical matrix element of (54) and verify each step carefully.

After this digression about unitary transformations of orthonormal sets of functions we go back to the Bloch functions $\psi(\kappa, \mathbf{r})$ and ask for a particular kind of unitary transformation, which will give us a set of functions with more localized properties that the Bloch functions. Wannier showed in 1937 that if we choose

$$V_{\kappa\mu} = \frac{1}{\sqrt{N}} \exp\left(-\frac{2\pi i \kappa \mu}{N}\right), \qquad (V.55)$$

we get a set of functions labeled by the cell subscripts μ:

$$W_\mu(\mathbf{r}) = \frac{1}{\sqrt{N}} \sum_{\kappa=-N/2}^{N/2-1} \psi(\kappa, \mathbf{r}) \exp\left(-\frac{2\pi i \kappa \mu}{N}\right). \quad (V.56)$$

96. Show explicitly that the matrix **V** (55) is unitary.

97. Specialize to the case $N = 4$, and write up the corresponding matrices explicitly.

98. Show that functions of type (56) associated with different cells are orthonormal.

In the previous derivations we have several times made use of two sum rules which we now "formalize." We get by direct summation of the geometric series

$$\sum_{\kappa = -N/2}^{N/2 - 1} \exp\left[\frac{2\pi i\kappa(\mu - \mu')}{N}\right] = N\delta_{\mu\mu'}; \qquad (V.57a)$$

$$\sum_{\mu = -N/2}^{N/2 - 1} \exp\left[\frac{2\pi i(\kappa - \kappa')\mu}{N}\right] = N\delta_{\kappa + \nu N, \kappa'}. \qquad (V.57b)$$

99. Verify (57) carefully.

100. Why are the two sums (57) slightly different? In other words, why is there nothing corresponding to νN in (57a)?

101. Illustrate (57) with the case $N = 4$ by marking the terms on the unit circle in the complex plane.

The Wannier functions (56) thus form an orthonormal set. There is one Wannier function associated with each cell. When the Bloch functions are of LCAO type, there is also one AO in each cell. AOs in different cells are, however, in general *not* orthogonal. But obviously there must exist some connection between the AOs and the Wannier functions that one can derive by means of (56) from the Bloch functions constructed from these AOs. That connection can be obtained from a combination of (29) and (56). It is instructive, however, to start out from an *un*normalized Bloch function [cf. (23) and (56)],

$$\psi'(\kappa, \mathbf{r}) = \mathbf{0}_\kappa \phi_0(\mathbf{r}) = \frac{1}{N} \sum_{\mu = -N/2}^{N/2 - 1} \phi_\mu(\mathbf{r}) e^{-2\pi i \kappa \mu / N}$$

$$= \frac{1}{\sqrt{N}} \sum_{\mu = -N/2}^{N/2 - 1} \phi_\mu(\mathbf{r}) (\mathbf{V}^+)_{\mu\kappa}. \qquad (V.58)$$

In terms of "fat symbols" the N relations (58) (why N?) can be written

$$\boldsymbol{\psi}' = \frac{1}{\sqrt{N}} \boldsymbol{\phi} \mathbf{V}^+. \qquad (V.59)$$

The normalization constants of the function (58) can be calculated as follows:

$$\boldsymbol{\psi}'^+ \boldsymbol{\psi}' = \frac{1}{N} \mathbf{V} \boldsymbol{\phi}^+ \boldsymbol{\phi} \mathbf{V}^+ = \frac{1}{N} \mathbf{V} \boldsymbol{\Delta} \mathbf{V}^+ = \frac{1}{N} \mathbf{d}. \qquad (V.60)$$

Here we have thus introduced the overlap matrix $\boldsymbol{\Delta}$ [cf. (25)], which is diagonalized by the unitary matrix $\mathbf{V}$.

EXERCISES

102. Calculate the normalization constant of *one* Bloch function (58).

103. Assume that the overlap integrals between the AOs decay so rapidly that it is sufficient to keep only those associated with nearest neighbors (and, of course, the normalization constant of the AO itself). Use the result of Exercise 102 to get an expression for the corresponding normalization constant of a Bloch function (58). Is that a real number? Should it?

104. Plot the normalization constant of a Bloch function (58) obtained in Exercise 103 as a function of κ for $-N/2 \leq \kappa < N/2$.

The overlap matrix (60) is *cyclic* [cf. (25)]:

$$\Delta_{\mu\nu} = \Delta_{\mu-p,\nu-p}, \tag{V.61}$$

which is why it is diagonalized by the particular matrix (55):

$$(\mathbf{V}\,\Delta\mathbf{V}^+)_{\kappa'\kappa} = \sum_{\mu,\nu}^{\text{BK}} V_{\kappa'\mu}\,\Delta_{\mu\nu}(\mathbf{V}^+)_{\nu\kappa}$$

$$= \frac{1}{N}\sum_{\mu,\nu}^{\text{BN}} \Delta_{\mu\nu}\,\exp\left[\frac{2\pi i(-\kappa'\mu + \nu\kappa)}{N}\right] = d_\kappa \delta_{\kappa'\kappa}. \tag{V.62}$$

The elements of the diagonal matrix **d** are given by

$$d_\kappa = \sum_\mu^{BK} \Delta_{\mu 0} \exp\left(-\frac{2\pi i \kappa \mu}{N}\right). \qquad \text{(V.63)}$$

EXERCISES

105. What is the reason for (61)? Is it true for all AOs?

106. Verify each step in (62). Notice that several steps are needed which are not spelled out explicitly.

107. What does (62) tell about **(a)** $\int dv\, \psi'^*(\kappa, \mathbf{r})\psi'(\kappa, \mathbf{r})$ and **(b)** $\int dv\, \psi'^*(\kappa', \mathbf{r})\psi'(\kappa, \mathbf{r})$, when $\kappa' \neq \kappa$?

108. What is the relation between d_k in (63) and N_κ in (28)?

The normalized Bloch function can then be written [does that agree with (29)?]

$$\psi'(\kappa, \mathbf{r}) = \sqrt{\frac{N}{d_\kappa}}\, \psi'(\kappa, \mathbf{r}), \qquad \text{(V.64)}$$

and the corresponding relation for the whole set of Bloch functions is

$$\psi = \sqrt{N}\, \psi' \mathbf{d}^{-1/2}. \qquad \text{(V.65)}$$

109. Why does the matrix $\mathbf{d}^{-1/2}$ in (65) have to be written to the right of ψ'?

110. Use (65) to show that $\psi^{+}\psi = \mathbf{1}$. What does that mean in words?

111. Combine (59) and (65) to express the final normalized Bloch functions in terms of the AOs.

Finally, we combine (65) with the definition of the Wannier functions (56) to get

$$\mathbf{W} = \psi\mathbf{V} = \phi\mathbf{V}^{+}\mathbf{d}^{-1/2}\mathbf{V}. \qquad (\text{V.66})$$

112. Use (66) to write *one* Wannier function explicitly.

113. Interpret the result of Exercise 112 to describe the Wannier function in terms of the AOs.

114. What is the matrix **d** when the AOs are orthonormal? What is then the relation between the Wannier functions and the AOs?

115. How does the LCAO expression (58) of an unnormalized Bloch function look when the AOs are orthonormal? Normalize it.

The theoretical tools described in this chapter can now be used, first, for a Hückel type of treatment analogous to Chapter IV. They are by no means limited to that level of approximation, however, but constitute a suitable starting point for almost all descriptions and calculations for polymers.

REFERENCES

André, J.-M., J. Delhalle, and J.-L. Brédas, *Quantum Chemistry Aided Design of Organic Polymers*, World Scientific Publishing Co., Singapore, 1991.

Brillouin, L., *Wave Propagation in Periodic Structures*, Dover, New York, 1953.

Slater, J. C., *Quantum Theory of Molecules and Solids*, Vol. II, McGraw-Hill, New York, 1965.

VI
CRYSTALS

VI.A. GENERALITIES

The polymers treated in Chapter V in a certain sense constitute an intermediate case between "finite systems" (atoms and small molecules) and extended (sometimes called infinite) systems. A polymer is extended in one dimension and "finite" in the other two dimensions. A surface or an interface is extended in two dimensions but behaves more or less like a small molecule in the third dimension.

All systems are obviously three-dimensional. If a system is "extended" in all three dimensions and if it is solid, we have basically two possibilities. Either the solid is disordered in one or several ways (structurally, chemically, magnetically, or other), or it has some kind of long-range order. In the latter case we talk about *crystals*. In the present chapter we discuss some fundamental aspects of the theory of crystals. Before delving into that subject it is worth noting, however, that for both ordered and disordered solids, periodic boundary conditions are called for to allow for wave functions to be nonzero throughout the system. Equally important is to keep in mind the necessity of working with a model that allows the long-range Coulomb forces to be operative throughout the system.

As in the other chapters of this book, we concentrate on the electronic structure in the Born–Oppenheimer approximation. We thus disregard the kinetic energy of the nuclei and study the electronic structure for a set of regularly spaced nuclear positions. From various kinds of diffraction experiments we know that crystals have long–range order. This means that the equilibrium positions of the nuclei form regular lattices. A long time before the crystallographers learned to evaluate crystal structures directly, the mathematicians had worked out the theory of space groups. We are not going to dwell on that fascinating subject in the present volume, but we note in passing that in three dimensions 230 different space groups are possible. Any crystal structure is thus invariant under all the transformations of one of those 230 groups. The number of possible lattice types is considerably smaller. There are 14 *Bravais lattices*. These provide a "structure" for the various combinations of translations that are possible. A space group is obtained when one of the Bravais lattices is combined with one of the 32 crystallographic point groups. (Why don't we get $14 \times 32 = 448$ space groups?)

We are thus going to study the electronic structure of a crystal with a lattice characterized by three basis vectors $\mathbf{a}_1$, $\mathbf{a}_2$, and $\mathbf{a}_3$. To qualify as basis vectors, they must not lie in the same plane, which is another way of saying that they must be linearly independent. The lattice vectors are linear combinations of the basis vectors with integer coefficients:

$$\mathbf{m} = \mathbf{a}_1\mu_1 + \mathbf{a}_2\mu_2 + \mathbf{a}_3\mu_3; \qquad \mu_i \text{ integer.} \qquad \text{(VI.1)}$$

We are going to impose boundary conditions like (V.2) in all three dimensions. It is therefore convenient to consider from the very beginning a Born–von Kármán (BK) region, which we define by restricting the integers μ_i in (1) to G possible values,

$$\text{BK: } -\frac{G}{2} \le \mu_i < \frac{G}{2}; \qquad i = 1, 2, 3. \qquad \text{(VI.2)}$$

Here the very large even integer G plays the same role as N in (V.2). This is a generalization to three dimensions of the cyclic condition used in Chapter V. Unfortunately, the three-dimensional situation does not allow for any simple geometric interpretation.

The boundary conditions that we impose on all wave functions for electrons in crystals are then

$$\psi(\mathbf{r} - G\mathbf{a}_i) = \psi(\mathbf{r}); \qquad i = 1, 2, 3. \qquad \text{(VI.3)}$$

These boundary conditions allow us to concentrate on the BK region characterized by (1) and (2). Surfaces and interfaces apparently require a separate treatment, but (3) is what is needed for all *bulk properties* of the crystals.

There are several important qualitative differences between a three-dimensional crystal and the polymers discussed in Chapter V. We illustrate this and some other essential aspects with reference to one of the three cubic lattices: simple cubic (sc), body-centered cubic (bcc), and face-centered cubic (fcc).

---

1. Define three basis vectors for a sc lattice with cube edge a.

2. List all the nearest neighbors of a particular lattice point in a sc lattice. How long is the distance to the nearest neighbor?

3. List all the next nearest neighbors of a particular lattice point in a sc lattice. How long is the distance to these lattice points?

4. Calculate the volume of the unit cell of a sc lattice.

5. Which is the volume of the BK region?

In the three-dimensional case it is natural to describe the system in terms of a central lattice point and neighbors of various orders. We must also keep track of the total number of lattice points and then it is practical to introduce the notation $N = G^3$ [cf. (2)].

In the three cubic Bravais lattices the cube obviously plays a central role. It is, however, only in the sc lattice that the cube

CRYSTALS

is also a *primitive unit cell*. By that term we mean a cell that contains only one lattice point. In the bcc and fcc lattices the primitive unit cells are not cubes, since the "fundamental cube" contains several lattice sites. The bcc lattice can be generated by superimposing two sc lattices in such a way that the sites of the second sc lattice are the centers of the cubes associated with the first one.

EXERCISES

6. How many nearest and next-nearest neighbors, respectively, does a lattice point of a bcc lattice have?

7. List all the first and second neighbors of a bcc lattice in terms of the three vectors $a\mathbf{e}_x$, $a\mathbf{e}_y$, and $a\mathbf{e}_z$.

8. How many lattice points in a bcc lattice are associated with the fundamental cube?

One possible choice of basis vectors for the bcc lattice is

$$\mathbf{a}_1 = \frac{a}{2}[-\mathbf{e}_x + \mathbf{e}_y + \mathbf{e}_z] = \frac{a}{2}(-1, 1, 1);$$

$$\mathbf{a}_2 = \frac{a}{2}[\mathbf{e}_x - \mathbf{e}_y + \mathbf{e}_z] = \frac{a}{2}(1, -1, 1);$$

$$\mathbf{a}_3 = \frac{a}{2}[\mathbf{e}_x + \mathbf{e}_y - \mathbf{e}_z] = \frac{a}{2}(1, 1, -1). \quad \text{(VI.4)}$$

Here we have introduced a shorthand notation to avoid writing out the unit vectors $\mathbf{e}_x$, $\mathbf{e}_y$, and $\mathbf{e}_z$.

170

9. Express the nearest neighbors of the lattice site 0 in a bcc lattice in terms of the basis vectors (4).

10. Calculate the nearest-neighbor distance from the result of Exercise 9 and check if this is consistent with earlier results.

For any lattice with a primitive unit cell generated by the basis vectors $\mathbf{a}_1$, $\mathbf{a}_2$, and $\mathbf{a}_3$, the volume of the primitive unit cell is given by

$$V_{0a} = \mathbf{a}_1 \cdot (\mathbf{a}_2 \times \mathbf{a}_3). \qquad \text{(VI.5)}$$

Since a volume must be a positive quantity, the expression (5) apparently presupposes that the labelling of the basis vectors has been made in a suitable way.

11. Use (5) to calculate the volume of the primitive cell for the sc lattice. What is the volume of BK in that case?

12. Use (5) to calculate the volume of the primitive cell for the bcc lattice. What is the volume of BK in that case?

171

CRYSTALS

So far we have not mentioned anything about the ''contents'' of the cells. The simplest (from this particular point of view) crystals are monatomic, which means that there is one atom per unit cell. Here we must distinguish carefully between lattice sites and atoms. A primitive unit cell always contains one lattice site, but there may be any number of atoms in that cell. Throughout the chapter we stick to the convention that BK contains $N = G^3$ primitive unit cells.

EXERCISES

13. How many electrons are there in a hypothetical sc hydrogen crystal?

14. Which is the average electron density in such a crystal?

15. How many electrons are there in a hypothetical bcc hydrogen crystal?

16. Which is the average electron density in bcc hydrogen crystal?

The sodium metal, like all the alkali metals under normal conditions, crystallizes in the bcc structure. Of the 11 electrons of each Na atom, one contributes to the ''sea'' of conduction electrons, which are delocalized throughout the crystal. The Ne-like ion core is more or less inert.

17. How many electrons does the BK region of a Na metal contain?

18. How many of those are conduction electrons?

19. What is the total electron density of the Na metal when the cube edge is $a = 4.23$ Å?

20. What, then, is the density of the conduction electrons?

VI.B. TRANSLATION SYMMETRY

The treatment of translation symmetry for a three-dimensional system with long-range order is a direct generalization of the corresponding section in Chapter V. We now have three translation operators, one for each direction.

$$\mathbf{T}_i \psi(\mathbf{r}) = \psi(\mathbf{r} - \mathbf{a}_i); \quad i = 1, 2, 3. \quad \text{(VI.6)}$$

EXERCISES

21. What are $\mathbf{T}_1\mathbf{T}_2\psi(\mathbf{r})$ and $\mathbf{T}_2\mathbf{T}_1\psi(\mathbf{r})$?

22. What is $\mathbf{T}_1^G$?

Since the three translation operators $\mathbf{T}_i$ commute,

$$[\mathbf{T}_i, \mathbf{T}_j] = 0; \quad i, j = 1, 2, 3, \qquad \text{(VI.7)}$$

they have simultaneous eigenfunctions. Their eigenvalues are given by (V.9).

$$\mathbf{T}_i \chi(\kappa_1, \kappa_2, \kappa_3; \mathbf{r}) = \exp\left(-\frac{2\pi i \kappa_i}{N}\right) \chi(\kappa_1, \kappa_2, \kappa_3; \mathbf{r});$$

$$-\frac{G}{2} \le \kappa_i < \frac{G}{2}; \quad i = 1, 2, 3.$$

$$\text{(VI.8)}$$

EXERCISES

23. Express the function $\mathbf{T}_1 \mathbf{T}_2 \chi(\kappa_1, \kappa_2, \kappa_3; \mathbf{r})$ in two different ways.

24. What is $\mathbf{T}_3^\mu \chi(\kappa_1, \kappa_2, \kappa_3; \mathbf{r})$, $\mathbf{T}_3^G \chi(\kappa_1, \kappa_2, \kappa_3; \mathbf{r})$?

25. Write $\mathbf{T}_1^{\mu_1} \mathbf{T}_2^{\mu_2} \mathbf{T}_3^{\mu_3} \chi(\kappa_1, \kappa_2, \kappa_3; \mathbf{r})$ both in terms of $\chi(\kappa_1, \kappa_2, \kappa_3; \mathbf{r})$ and using the vector $\mathbf{m}$ in (1).

The notation used so far is obviously not very practical. The form of the eigenvalue corresponding to the operator product $\mathbf{T}_1^{\mu_1} \mathbf{T}_2^{\mu_2} \mathbf{T}_3^{\mu_3}$,

$$\exp\left\{-\frac{2\pi i}{G}(\kappa_1\mu_1 + \kappa_2\mu_2 + \kappa_3\mu_3)\right\}, \qquad (VI.9)$$

is reminiscent of a scalar product. The vector in the direct lattice,

$$\mathbf{m} = \mathbf{a}_1\mu_1 + \mathbf{a}_2\mu_2 + \mathbf{a}_3\mu_3, \qquad (VI.10)$$

multiplied by a so far unknown vector is supposed to yield the expression in braces in (9). We recall that in general the three basis vectors $\mathbf{a}_i$ are not perpendicular to each other.

EXERCISES

26. With the notation $\mathbf{a}_i = (\mathbf{a}_{ix}, \mathbf{a}_{iy}, \mathbf{a}_{iz}) = \mathbf{e}_x\mathbf{a}_{ix} + \mathbf{e}_y\mathbf{a}_{iy} + \mathbf{e}_z\mathbf{a}_{iz}$, what is the scalar product $\mathbf{a}_1 \cdot \mathbf{a}_2$?

27. What is the length of the vector $\mathbf{a}_i$?

28. Is the set of basis vectors $\{\mathbf{a}_1, \mathbf{a}_2, \mathbf{a}_3\}$ orthonormal for **(a)** sc; **(b)** bcc?

Given a set of linearly independent vectors $\mathbf{a}_i$, even if they are not orthogonal, we can construct the corresponding *biorthogonal set* [cf. (5)]

$$\mathbf{b}_1 = \frac{\mathbf{a}_2 \times \mathbf{a}_3}{V_{0a}};$$

$$\mathbf{b}_2 = \frac{\mathbf{a}_3 \times \mathbf{a}_1}{V_{0a}};$$

$$\mathbf{b}_3 = \frac{\mathbf{a}_1 \times \mathbf{a}_2}{V_{0a}}. \qquad (VI.11)$$

EXERCISES

29. Calculate the three vectors $\mathbf{b}_j$ for the sc and bcc lattices.

30. Calculate both for sc and bcc: $\mathbf{a}_1 \cdot \mathbf{b}_1; \mathbf{a}_1 \cdot \mathbf{b}_2; \mathbf{a}_1 \cdot \mathbf{b}_3.$

31. Calculate in both cases $\mathbf{b}_1 \cdot (\mathbf{b}_2 \times \mathbf{b}_3).$

Using (5) and (11), one can show quite generally (do that!) that

$$\mathbf{a}_i \cdot \mathbf{b}_j = \delta_{ij}; \qquad i, j = 1, 2, 3. \qquad (VI.12)$$

We define a certain linear combination of the $\mathbf{b}_j$:

$$\mathbf{k} = \frac{2\pi}{G}(\mathbf{b}_1 \kappa_1 + \mathbf{b}_2 \kappa_2 + \mathbf{b}_3 \kappa_3);$$

$$-\frac{G}{2} \le \kappa_i < \frac{G}{2}; i = 1, 2, 3. \qquad (VI.13)$$

With (10), (12), and (13) we then get

$$\mathbf{k \cdot m} = \frac{2\pi}{G} (\kappa_1 \mu_1 + \kappa_2 \mu_2 + \kappa_3 \mu_3), \qquad \text{(VI.14)}$$

which implies that the eigenvalue (9) can be written

$$\exp(-i\mathbf{k \cdot m}). \qquad \text{(VI.15)}$$

Going back to (8) we can thus write the general eigenvalue relation for the translation operators as

$$\mathbf{T}(\mathbf{m})\chi(\mathbf{k}; \mathbf{r}) = \chi(\mathbf{k}; \mathbf{r}) \, e^{-i\mathbf{k \cdot m}}. \qquad \text{(VI.16)}$$

Here [cf. (10)]

$$\mathbf{T}(\mathbf{m}) = \mathbf{T}_1^{\mu_1}\mathbf{T}_2^{\mu_2}\mathbf{T}_3^{\mu_3}. \qquad \text{(VI.17)}$$

It is important to realize that (16) is a condensed way of expressing three mutually compatible eigenvalue relations. The *wave vector* $\mathbf{k}$, (13), labels both the eigenvalue, $\exp(-i\mathbf{k \cdot m})$, and the eigenfunction, $\chi(\mathbf{k}; \mathbf{r})$, of $\mathbf{T}(\mathbf{m})$.

EXERCISES

32. What is the value of the eigenvalue $\exp(-i\mathbf{k \cdot m})$, with the translation operator characterized by the lattice vector $\mathbf{m} = \mathbf{a}_1$, when (a) $\mathbf{k} = 0$; (b) $\kappa_1 = 1$; $\kappa_2 = \kappa_3 = 0$; (c) $\kappa_1 = -G/2$; $\kappa_2 = \kappa_3 = 0$; and (d) $\kappa_1 = \kappa_2 = \kappa_3 = -G/2$?

33. What are the corresponding eigenvalues when $\mathbf{m} = \mathbf{a}_1 - \mathbf{a}_2 + \mathbf{a}_3$?

Since each one of the integers κ_i can take G different values, there are $N = G^3$ different wave vectors $\mathbf{k}$.

---

34. Use the results of Exercise 29 to give an expression for a typical wave vector **k** in the bcc case, in terms of the three unit vectors $\mathbf{e}_x$, $\mathbf{e}_y$, $\mathbf{e}_z$,

35. What is the length $k = |\mathbf{k}|$ of such a wave vector? Which dimension does a wave vector have?

36. Find all wave vectors **k** in the bcc case of length $2\pi\sqrt{6}/Ga$.

The $N = G^3$ different values of **k** thus label the different symmetry types with respect to translations. This is nothing but a generalization of the two possibilities for the MO of a hydrogen molecule, (IV.28), or an ethylene molecule, (IV.60), the six possibilities for benzene, (IV.73), or a "polymer" with only two links (V.29).

Just like the basis vectors $\mathbf{a}_i$ the vectors $\mathbf{b}_j$ can be used to generate a lattice—the *reciprocal lattice*. A typical reciprocal lattice vector is then

$$\mathbf{K} = 2\pi(\mathbf{b}_1 \, \nu_2 + \mathbf{b}_2\nu_2 + \mathbf{b}_3\nu_3);$$

$$\nu_j \text{ any positive or negative integer or zero.} \quad \text{(VI.18)}$$

The factor 2π is included to get a notation consistent with (13). We notice that a capital **K** is used for the lattice vector (18), whereas a lowercase **k** is used for (13), which is in general not a lattice vector.

37. For which values of κ_i are the wave vectors **k** reciprocal lattice vectors?

38. Is there any relation between some reciprocal lattice vector and a wave vector **k** with $\kappa_1 = -G/2$; $\kappa_2 = \kappa_3 = 0$?

39. List the nearest-neighbor reciprocal lattice vectors in the reciprocal lattice associated with a bcc direct lattice.

The term *reciprocal* is obviously connected with (12). From (11) we calculate the volume of the unit cell formed by the three basis vectors $\mathbf{b}_j$:

$$V_{0b} = \mathbf{b}_1 \cdot (\mathbf{b}_2 \times \mathbf{b}_3) = \frac{1}{V_{0a}}. \qquad (VI.19)$$

The unit cell volume in the reciprocal lattice is the reciprocal of that in the direct lattice. The periodic boundary conditions imply that we work with a *finite direct lattice* [cf. (2)]. For the reciprocal lattice there is no corresponding limitation, however: the reciprocal lattice is infinite.

It is important to understand the different roles played by the two types of lattice or space. In *position space* we use **r** to label the various points and the wave functions in the position representation are functions of **r**. The *direct lattice* consists of a large but finite number of points **m** of type (10). The crystal (the BK region) is invariant under all translations associated with the direct lattice vectors **m**, and it is therefore sensible to work with eigenfunctions of the translation operators. The $N = G^3$ wave vectors **k**, (13), label the possible symmetry types with respect to translations. These wave vectors lie in *reciprocal space*, which also contains the infinitely many, but discretely labeled *reciprocal lattice vectors* **K**, (18). As we will see later, there is an intimate connection between the discrete wave vectors **k** and the discretization of *momentum space*, which is associated with the periodic boundary conditions [cf. (V.34)].

EXERCISES

40. What is the scalar product of a reciprocal and a direct lattice vector?

41. What is then $\exp(i\mathbf{K}\cdot\mathbf{m})$?

42. What is $\exp(i(\mathbf{k} + \mathbf{K})\cdot\mathbf{m})$?

As we have seen [cf. (V.8) and (9)] there are only $N = G^3$ different eigenvalues of the translation operators. The wave vectors **k** and **k** + **K**, where **K** is any reciprocal lattice vector, therefore refer to the same eigenvalue (9). The N vectors **k** defined by (13) are thus sufficient for the characterization of the eigenfunctions and eigenvalues of the translation operators. They define the so-called *first Brillouin zone* [BZ], which is a term used for a unit cell in the reciprocal lattice centered around a lattice point. Its bordering planes are obtained by setting $\kappa_1 = -G/2$; $\kappa_2 = \kappa_3 = 0$, which gives $\mathbf{k} = 2\pi\cdot(\mathbf{b}_1/2)$, and similarly in the other directions. Comparing with (18) we see that these bordering planes are the planes that bisect the reciprocal lattice vectors perpendicularly. A simple illustration of the principles involved can be obtained if that procedure is applied to a planar square lattice.

EXERCISES

43. To what does BZ correspond in the one-dimensional case treated in Chapter V?

44. Which are the counterparts in the one-dimensional case to the wave vectors **k** and the reciprocal lattice vectors **K**?

All the results obtained for a one-dimensional lattice in Chapter V can be generalized to three dimensions. In particular we can use the sum rules (V.57) in each one of the three dimensions to get

$$\sum_{\mathbf{k}}^{BZ} e^{i\mathbf{k}\cdot(\mathbf{m'}-\mathbf{m})} = N\delta_{\mathbf{m'},\mathbf{m}}; \qquad (VI.20a)$$

$$\sum_{\mathbf{m}}^{BK} e^{i(\mathbf{k'}-\mathbf{k})\cdot\mathbf{m}} = N\delta_{\mathbf{k'},\mathbf{k}+\mathbf{K}}. \qquad (VI.20b)$$

EXERCISES

45. Why is there a difference in the structure of the two sum rules (20a) and (20b)?

46. What does the notation for the two sum rules mean explicitly?

In the one-dimensional case we generated eigen*functions* of the translation operators from arbitrary functions by means of projection operators. That procedure can be taken over directly to three dimensions, both with the sum and the product form of the projection operators. The three-dimensional sum form corresponding to (V.20) is

$$\mathbf{O_k} = \frac{1}{N} \sum_{\mathbf{m}}^{\text{BK}} e^{i\mathbf{k}\cdot\mathbf{m}} \, \mathbf{T}\,(\mathbf{m}). \qquad (VI.21)$$

EXERCISES

47. What does the notation for the three sums in (21) stand for explicitly? Notice that $N = G^3$.

48. Show that $\mathbf{O_{k'}}, \mathbf{O_k} = \delta_{\mathbf{k'},\mathbf{k}} \mathbf{O_k}$.

49. Show that $\sum_{\mathbf{k}}^{\text{BZ}} \mathbf{O_k} = \mathbf{1}$ (the identity operator).

For the application of (21) it is practical to introduce the notation

$$\mathbf{T}(\mathbf{m})\phi(\mathbf{r}) = \phi(\mathbf{r} - \mathbf{m}) = \phi(\mathbf{m}; \mathbf{r}). \qquad (VI.22)$$

50. What is $\mathbf{T(m)}\phi(\mathbf{r})$ explicitly when $\phi(\mathbf{r})$ is a hydrogen atom $1s$ orbital and $\mathbf{m} = \mathbf{e}_x + \mathbf{e}_y + \mathbf{e}_z$?

51. What is $\mathbf{T(m)}\,\mathbf{T(n)}\phi(\mathbf{r})$ and $\mathbf{T(n)}\,\mathbf{T(m)}\phi(\mathbf{r})$, respectively?

52. $\mathbf{T(m)}$ is a short way of writing a product of three translation operators [cf.(17)]. What is $\mathbf{T(m)}\phi(\mathbf{r})$ when $\mu_1 = G$; $\mu_2 = 3G$; $\mu_3 = 5G$?

We then apply (21) to an arbitrary function $\phi(\mathbf{r})$:

$$\mathbf{O}_k\phi(\mathbf{r}) = \frac{1}{N} \sum_{\mathbf{m}}^{\text{BK}} e^{i\mathbf{k}\cdot\mathbf{m}}\, \phi(\mathbf{m}; \mathbf{r}). \qquad \text{(VI.23)}$$

53. Show that (23) is an eigenfunction of $\mathbf{T(n)}$ for any direct lattice vector $\mathbf{n}$. Which is the corresponding eigenvalue?

54. In Exercise 53 a relabeling of the terms is necessary. Motivate and show how and why.

55. What is obtained when $\mathbf{O}_{\mathbf{k}'}$ operates on (23)? The result can be obtained in several different ways.

Two Bloch functions are orthogonal when they are labeled by different wave vectors in BZ. The simplest way to see that is to use the generalization of Exercise V.57, together with the turn over rule:

$$\int dv \, [\mathbf{O}_{\mathbf{k}'}\phi(\mathbf{r})]^* \, \mathbf{O}_{\mathbf{k}} \, \phi(\mathbf{r}) = \int dv \, \phi^*(\mathbf{r}) \, \mathbf{O}_{\mathbf{k}'}^+, \, \mathbf{O}_{\mathbf{k}} \, \phi(\mathbf{r})$$

$$= \int dv \, \phi^*(\mathbf{r}) \, \mathbf{O}_{\mathbf{k}'} \mathbf{O}_{\mathbf{k}} \, \phi(\mathbf{r})$$

$$= \delta_{\mathbf{k}',\mathbf{k}} \int dv \, \phi^*(\mathbf{r}) \, \mathbf{O}_{\mathbf{k}} \, \phi(\mathbf{r}).$$

$$(VI.24)$$

EXERCISES

56. To carry out the various steps in (24), one must know that $\mathbf{O}_{\mathbf{k}}$ is a Hermitian operator. Why does it have that property even though the translation operators are *not* Hermitian?

57. What happens in (24) if we use two different "starting functions"? Interpret the result?

For the same reason as in Chapter V, the projected function (23) is in general (does it ever happen?) not normalized when

$\phi(\mathbf{r})$ is normalized. With the notation

$$\Delta(\mathbf{m}, \mathbf{n}) = \int dv \, \phi^*(\mathbf{m}; \mathbf{r}) \phi(\mathbf{n}; \mathbf{r}) \qquad \text{(VI.25)}$$

for the overlap integral between the functions centered at (or rather labeled by) $\mathbf{m}$ and $\mathbf{n}$, we have for the normalization constant of (23),

$$\int dv \, [\mathbf{O}_k \phi(\mathbf{r})]^* \, \mathbf{O}_k \phi(\mathbf{r}) = \frac{1}{N} \sum_{\mathbf{m}}^{\text{BK}} \Delta(\mathbf{O}, \mathbf{m}) e^{i\mathbf{k} \cdot \mathbf{m}} = N(\mathbf{k}).$$

$$\text{(VI.26)}$$

EXERCISES

58. Carry out (26) explicitly and motivate each step.

59. What is $N(\mathbf{k})$ if the functions $\phi(\mathbf{m}; \mathbf{r})$ and $\phi(\mathbf{n}; \mathbf{r})$ are orthonormal?

60. What is $N(\mathbf{k})$ for an sc lattice if only the nearest-neighbor overlap integrals are taken into account? How can the expression be simplified if the overlap integrals depend only on the distance between $\mathbf{m}$ and the origin?

The discussion presented so far is directly applicable to the case when $\phi(\mathbf{r})$ is an AO (i.e., has an appreciable amplitude only in the neighborhood of the origin). For crystals there is another type of function that is just as important, namely the plane waves (PW),

$$\eta(\mathbf{k};\,\mathbf{r}) = \frac{1}{\sqrt{V}}\,e^{i\mathbf{k}\cdot\mathbf{r}}. \qquad\qquad (\text{VI.27})$$

Here $V = NV_{0a}$ is the total volume of BK.

EXERCISES

61. Show that the PW defined in (27) is normalized in BK.

62. Show that $\mathbf{O}_{\mathbf{k}'}\,\eta(\mathbf{k};\,\mathbf{r}) = \delta_{\mathbf{k}',\,\mathbf{k}+\mathbf{K}}\eta(\mathbf{k};\,\mathbf{r})$.

63. What is $\mathbf{O}_{\mathbf{k}}\eta(\mathbf{k}+\mathbf{K};\,\mathbf{r})$?

64. Is (27) a Bloch function?

It should be noted that a wave vector $\mathbf{k}$ which characterizes the plane wave (27) can be anywhere in reciprocal space—it is by no means restricted to BZ. But as seen in the exercises, the symmetry type with respect to translations is labeled by one of the wave vectors in BZ.

We can see directly that two PWs with different wave vectors are orthogonal in BK. We write $\mathbf{r} = \mathbf{a}_1\rho_1 + \mathbf{a}_2\rho_2 + \mathbf{a}_3\rho_3$ and $\mathbf{k} = \mathbf{b}_1 k_1 + \mathbf{a}_2 k_2 + \mathbf{a}_3 k_3$, and get

$$\int dv \, \eta^*(\mathbf{k}'; \mathbf{r})\eta(\mathbf{k}; \mathbf{r}) = \frac{1}{V} \int dv \, e^{i(\mathbf{k}-\mathbf{k}')\cdot\mathbf{r}}$$

$$= \frac{V_{0a}}{V} \prod_{j=1}^{3} \int_{-G/2}^{G/2} d\rho_j \, e^{i(k_j - k_j')\rho_j}$$

$$= \frac{1}{N} \prod_{j=1}^{3} \frac{G \sin \pi(\kappa_j - \kappa_j')}{\pi(\kappa_j - \kappa_j')}.$$

$$(VI.28)$$

This product vanishes unless $\kappa_j = \kappa_j'$ for $j = 1, 2, 3$. For $\mathbf{k}' = \mathbf{k}$ (28) shows that $\eta(\mathbf{k}; \mathbf{r})$ is normalized.

EXERCISES

65. What is the origin of the factor V_{0a} in (28)?

66. Calculate directly the overlap integral between the two PWs $\eta(\mathbf{k}; \mathbf{r})$ and $\eta(k + \mathbf{K}; \mathbf{r})$. Relate the result to (28).

It is very instructive to plot a few PWs corresponding to different wave vectors as functions of $\mathbf{r}$ for one of the common lattices. Choose a sc lattice with $\mathbf{b}_1 = (1/a) \, \mathbf{e}_x$; $\mathbf{b}_2 = (1/a) \, \mathbf{e}_y$; $\mathbf{b}_3 = (1/a) \, \mathbf{e}_z$.

EXERCISES

67. Write up explicitly the wave vectors $\mathbf{k}$ for the following sets of integers κ_j: **(a)** $\kappa_1 = \kappa_2 = \kappa_3 = 0$; **(b)** $\kappa_1 = 1$; $\kappa_2 = \kappa_3 = 0$; $\kappa_1 = 0$; $\kappa_2 = \kappa_3 = 1$; **(d)** $\kappa_1 = -G/2$; $\kappa_2 = \kappa_3 = 0$.

68. Plot the PWs in these four cases as functions of x, y, z or some of these variables. Calculate their wavelengths and relate these to the corresponding wave vectors.

VI.C. MOMENTUM SPACE

The momentum-space counterpart of a function satisfying the periodic boundary conditions (3) satisfies the condition [cf. (V.32)]

$$\underline{\phi}(\mathbf{p}) = \frac{1}{\sqrt{8\pi^3}} \int dv\, \phi(\mathbf{r} - G\mathbf{a}_j)e^{-i\mathbf{p}\cdot\mathbf{r}}$$

$$= \frac{1}{\sqrt{8\pi^3}} \int dv'\, \phi(\mathbf{r}')e^{-i\mathbf{p}\cdot(\mathbf{r}' + G\mathbf{a}_j)} = \underline{\phi}(\mathbf{p})e^{-i\mathbf{p}\cdot G\mathbf{a}_j};$$

$$j = 1, 2, 3. \tag{VI.29}$$

This implies that $\underline{\phi}(\mathbf{p})$—the momentum space counterpart of $\phi(\mathbf{r})$—vanishes unless

$$e^{-i\mathbf{p}\cdot G\mathbf{a}_j} = 1; \quad j = 1, 2, 3. \tag{VI.30}$$

If we write the momentum variable in terms of the reciprocal basis (11),

$$\mathbf{p} = \mathbf{b}_1 p_1 + \mathbf{b}_2 p_2 + \mathbf{b}_3 p_3, \tag{VI.31}$$

the conditions (30) can be written

$$\exp(-iGp_j) = 1; \quad j = 1, 2, 3. \tag{VI.32}$$

This means that the components of the momentum along the reciprocal basis vectors must be of the form

$$p_j = \frac{2\pi\lambda_i}{G}; \quad \lambda_j \text{ positive or negative integer or zero.}$$

$$\tag{VI.33}$$

EXERCISES

69. Are the integers in (33) limited in any way?

70. Compare (33) and (13). Why are the κ_i limited?

71. What is $\underline{\phi}(\mathbf{p})$ for $p_1 = p_2 = 0; p_3 = \pi/G$?

Thus for an extended system with periodic boundary cond-tions in three directions in position space, the momentum space is in a certain sense *discretized* in the corresponding directions.

In calculating the momentum-space counterpart of a function satisfying (3), it is sufficient to carry out the integration in (III.5) over the BK region. For a PW this means

$$\underline{\eta}(\mathbf{k}; \mathbf{p}) = \frac{1}{\sqrt{8\pi^3}} \int_{\text{BK}} dv\, \eta(\mathbf{k}; \mathbf{r}) e^{-i\mathbf{p}\cdot\mathbf{r}}$$

$$= \frac{1}{\sqrt{8\pi^3}\, V} \int_{\text{BK}} dv\, e^{i(\mathbf{k} - \mathbf{p})\cdot\mathbf{r}}. \qquad (\text{VI.34})$$

EXERCISES

72. Write $\mathbf{r} = \mathbf{a}_1\rho_1 + \mathbf{a}_2\rho_2 + \mathbf{a}_3\rho_3$ and carry out the integration in (34). Notice the Jacobian, which is needed explicitly since the basis vectors are in general not orthogonal.

73. What would the corresponding integral have given if the vectors **k** and **p** had been continuous variables and the BK region infinite?

The momentum-space counterpart of a plane wave is thus

$$\underline{\eta}(\mathbf{k};\, \mathbf{p}) = \sqrt{\frac{V}{8\pi^3}}\, \delta_{\mathbf{k},\mathbf{p}}. \qquad\qquad \text{(VI.35)}$$

The fact that we get such a strange "function" of **p** is an aspect of the discretization of momentum space. In (35) it is important to notice that **p** is the variable, whereas **k** is a "label." The function (35) vanishes for all **p** except when the variable **p** is the same as the label **k**.

EXERCISES

74. Does the plane wave (27) satisfy the periodic boundary conditions (3)?

75. Compare the momentum-space counterpart of an arbitrary function satisfying (3) with (35). How does the specialization to a plane wave show up?

76. Is there any relation between the momentum-space counterparts of PWs labeled **k** and **k** + **K**, where **K** is a reciprocal lattice vector?

A plane wave represents one extreme form of Bloch function. At the other conceptual extreme we find the Bloch sums generated from atomic orbitals, for example, by means of the projection operators (21),

$$\psi(\mathbf{k};\,\mathbf{r}) = \frac{1}{\sqrt{N(\mathbf{k})}}\, \mathbf{O_k}\phi(\mathbf{r})$$

$$= \frac{1}{N\sqrt{N(\mathbf{k})}}\, \sum_{\mathbf{m}}^{\mathrm{BK}} e^{i\mathbf{k}\cdot\mathbf{m}}\phi(\mathbf{m};\,\mathbf{r}). \qquad (\mathrm{VI.36})$$

We recall that [cf. (V.35)] a translation in position space corresponds to a phase factor in momentum space:

$$\frac{1}{\sqrt{8\pi^3}} \int_{\mathrm{BK}} dv\,\phi(\mathbf{m};\,\mathbf{r})\,e^{-i\mathbf{p}\cdot\mathbf{r}} = \underline{\phi}(\mathbf{p})e^{-i\mathbf{p}\cdot\mathbf{m}}. \qquad (\mathrm{VI.37})$$

EXERCISES

77. Verify (37).

78. Does it matter whether the integration in (37) is carried out over BK or over all space when $\phi(\mathbf{r})$ is an AO?

79. Is there any value of $\mathbf{p}$, which reduces the exponential factor in (37) to 1?

We use (37) to calculate the momentum space counterpart of the Bloch sum (36):

$$\underline{\psi}(\mathbf{k}; \mathbf{p}) = \frac{\phi(\mathbf{p})}{N\sqrt{N(\mathbf{k})}} \sum_{\mathbf{m}}^{BK} e^{i(\mathbf{k} - \mathbf{p}) \cdot \mathbf{m}}. \qquad (VI.38)$$

EXERCISES

80. The summation in (38) can be carried out explicitly. What does it give?

81. The result of Exercise 80 shows that an LCAO Bloch sum in a certain sense is intermediate between a PW and a general function satisfying periodic boundary conditions. Which sense?

82. Is there any relation between $\underline{\psi}(\mathbf{k}; \mathbf{p})$ and $\underline{\psi}(\mathbf{k} + \mathbf{K}; \mathbf{p})$?

We see from (38) two characteristic aspects of the momentum-space counterpart of an LCAO Bloch sum. On the one hand, as the counterpart of a function satisfying periodic boundary conditions, it vanishes unless the components of $\mathbf{p}$ satisfy (33). But as the counterpart of a Bloch sum, it vanishes unless the momentum variable $\mathbf{p}$ is equal to the wave vector $\mathbf{k}$ plus a possible reciprocal lattice vector $\mathbf{K}$. In other workds, for a given "label" $\mathbf{k}$ the variable $\mathbf{p}$ must differ from $\mathbf{k}$ by a reciprocal lattice vector $\mathbf{K}$, in order for the function to be different from zero.

VI.D. FROM INTEGRATION TO SUMMATION IN MOMENTUM SPACE

The discretization of momentum space discussed earlier implies that all integrals over functions of the momentum variable $\mathbf{p}$ should be replaced by sums. Then we must keep track of the density of points in momentum space. That density is uniform and we have, for example, from (13), that it is

$$\frac{N}{8\pi^3 V_{0b}} = \frac{V}{8\pi^3}. \qquad (VI.39)$$

EXERCISES

83. What is the origin of the factor $8\pi^3$ in (39)?

84. Does (39) check with (13) and (33)?

85. What happens with the density in momentum space when BK becomes very large? Is the answer to that question reasonable?

A volume element $d\mathbf{p}$ in momentum space contains $(V/8\pi^3)$ $d\mathbf{p}$ discrete points $\mathbf{p}$. We therefore have the following correspondence between sums and integrals in momentum space:

$$\int d\mathbf{p}\, f(\mathbf{p}) \leftrightarrow \frac{8\pi^3}{V} \sum_{\mathbf{p}} f(\mathbf{p}). \qquad (VI.40a)$$

By writing this expression instead as

$$\frac{V_{0a}}{8\pi^3} \int d\mathbf{p}\, f(\mathbf{p}) \leftrightarrow \frac{1}{N} \sum_{\mathbf{p}} f(\mathbf{p}), \qquad (VI.40b)$$

we see that the finite integer N associated with the periodic boundary conditions is "just" an auxiliary quantity.

86. Carry out (40) with $f(\mathbf{p}) = 1$ over BZ. Do the two results agree?

87. We should thus in principle calculate expressions of type (40) as sums. Sometimes, however, it might be easier to carry out the integral to the left than the sum to the right. Is it defensible to do so?

When we work with position-space functions satisfying periodic boundary conditions, the general transformation formula (III.5) for the Fourier transform connecting position and momentum space functions should thus be replaced by the sum

$$\phi(\mathbf{r}) = \frac{\sqrt{8\pi^3}}{V} \sum_{\mathbf{p}}^{\text{all}} \underline{\phi}(\mathbf{p})\, e^{i\mathbf{p}\cdot\mathbf{r}}, \qquad \text{(VI.41)}$$

in which we sum over all those values of $\mathbf{p}$ for which the function $\underline{\phi}(\mathbf{p})$ is different from zero, [cf. (33)].

88. Use (41) to ''resuscitate'' the plane wave from the momentum-space counterpart (35).

89. Substitute (41) in (III.5) and carry out the integration. Is the result consistent with what we should expect?

Then we apply (41) to the function (38),

$$\frac{\sqrt{8\pi^3}}{V} \sum_{\mathbf{p}}^{\text{all}} \underline{\psi}(\mathbf{k}; \mathbf{p}) e^{i\mathbf{p}\cdot\mathbf{r}} = \frac{1}{V}\sqrt{\frac{8\pi^3}{N(\mathbf{k})}} \sum_{\mathbf{K}}^{\text{all}} \underline{\phi}(\mathbf{k} + \mathbf{K}) e^{i(\mathbf{k} + \mathbf{K})\cdot\mathbf{r}}.$$

(VI.42)

What we have derived here is a plane-wave expansion of the Bloch sum $\psi(\mathbf{k}; \mathbf{r})$. Apart from certain constants, the coefficients of the PWs are apparently the values of the momentum-space counterparts of the basic AO at $\mathbf{p} = \mathbf{k} + \mathbf{K}$.

EXERCISES

90. The expansion (42) contains infinitely many terms but not all possible PWs. Which are missing?

91. The function $\exp(i\mathbf{k}\cdot\mathbf{r})$ can be factored out of (42). What characterizes the remaining factor?

92. Is the answer to Exercise 91 consistent with the fact that the remaining factor only contains PWs labeled by reciprocal lattice vectors?

In the last few exercises we have encountered a property common to all Bloch functions in position space: any Bloch function can be written as a product of a PW and a function that has the periodicity of the lattice. To see that we just write a Bloch function $\chi(\mathbf{k}; \mathbf{r})$, thus satisfying (16), as a plane wave with the same wave vector as the Bloch function, and a remaining factor $u(\mathbf{k}; \mathbf{r})$:

$$\chi(\mathbf{k}; \mathbf{r}) = e^{i\mathbf{k}\cdot\mathbf{r}}u(\mathbf{k}; \mathbf{r}). \tag{VI.43}$$

The function $u(\mathbf{k}; \mathbf{r})$ then transforms under translations as follows:

$$\mathsf{T}(\mathbf{m})u(\mathbf{k}; \mathbf{r}) = u(\mathbf{k}; \mathbf{r} - \mathbf{m})$$

$$= e^{-i\mathbf{k}\cdot(\mathbf{r} - \mathbf{m})}\chi(\mathbf{k}; \mathbf{r} - \mathbf{m}) = u(\mathbf{k}; \mathbf{r}), \tag{VI.44}$$

which shows that it has the same periodicity as the direct lattice.

EXERCISES

93. Does a Bloch function in general have the same periodicity as the lattice?

94. How does the answer to Exercise 93 agree with the fundamental periodic boundary conditions?

95. Is there any Bloch function that has the same periodicity as the lattice?

96. Which periodic function $u(\mathbf{k}; \mathbf{r})$ is associated with a plane wave?

The PWs (27) form a complete set, which means that any "reasonable" function can be expanded in such a set. One way to see that is to start out from the relation [cf. (40) and (II.6)],

$$\sum_{\mathbf{k}}^{\text{all}} \eta(\mathbf{k}; \mathbf{r})\eta^*(\mathbf{k}; \mathbf{r}') = \frac{1}{V} \sum_{\mathbf{k}}^{\text{all}} e^{i\mathbf{k}\cdot(\mathbf{r} - \mathbf{r}')}$$

$$\leftrightarrow \frac{1}{8\pi^3} \int d\mathbf{k}\, e^{i\mathbf{k}\cdot(\mathbf{r} - \mathbf{r}')} = \delta(\mathbf{r} - \mathbf{r}'), \quad (\text{VI.45})$$

which constitutes one definition of a complete set.

EXERCISES

97. Why does $\mathbf{k}$ in (45) play a role similar to that of $\mathbf{p}$ in (40)?

98. Try to interpret (45) explicitly.

99. What does "all" in the sum in (45) mean explicitly?

A PW expansion of a function $f(\mathbf{r})$ can then be obtained as follows:

$$f(\mathbf{r}) = \int dv'\, \delta(\mathbf{r} - \mathbf{r}')f(\mathbf{r}')$$

$$= \sum_{\mathbf{k}}^{\text{all}} \eta(\mathbf{k}; \mathbf{r}) \int dv'\, \eta^*((\mathbf{k}; \mathbf{r}')f(\mathbf{r}') = \sum_{\mathbf{k}}^{\text{all}} \eta(\mathbf{k}; \mathbf{r})f_k.$$

$$(\text{VI.46})$$

100. Which is the explicit expression for the coefficient f_k of the plane wave $\eta(\mathbf{k}; \mathbf{r})$ in the expansion (46)?

101. Calculate the coefficient f_k for a function that is constant.

102. Use (35) to get an expression for the momentum-space counterpart of (46).

103. Interpret the result of Exercise 102. What is the relation between the momentum-space counterpart of a function and its expansion in plane waves?

If the function $f(\mathbf{r})$ in (46) has the periodicity of the direct lattice, that is, if

$$f(\mathbf{r}) = f(\mathbf{r} - \mathbf{m}) \qquad \text{for all } \mathbf{m} \text{ in BK}, \qquad \text{(VI.47)}$$

the corresponding coefficients satisfy

$$f_k = \int dv\, \eta^*(\mathbf{k}; \mathbf{r}) f(\mathbf{r}) = \frac{1}{\sqrt{V}} \int dv\, f(\mathbf{r}) e^{-i\mathbf{k}\cdot\mathbf{r}}$$

$$= \frac{1}{\sqrt{V}} \int dv\, f(\mathbf{r} - \mathbf{m}) e^{-i\mathbf{k}\cdot\mathbf{r}} = \frac{1}{\sqrt{V}} \int dv'\, f(\mathbf{r}') e^{-i\mathbf{k}\cdot(\mathbf{r}' + \mathbf{m})}$$

$$= f_k\, e^{-i\mathbf{k}\cdot\mathbf{m}}. \qquad \text{(VI.48)}$$

104. What does (48) imply for f_k?

105. Is the result of Exercise 104 consistent with Exercise 101 when $f(\mathbf{r})$ is a constant?

As for the one-dimensional lattice treated in Chapter V, we can define Wannier functions, which can be made more localized than the Bloch functions. It is a good exercise to carry through that transformation in the three-dimensional case. The theoretical "tools" developed in this chapter are needed for all kinds of calculations of energies and other properties of crystals.

REFERENCES

Altmann, S. L., *Band Theory of Metals*: *The Elements*, Pergamon Press, Oxford, 1970.

Brillouin, L., *Wave Propagation in Periodic Structures*, Dover, New York, 1953.

Löwdin, P.-O. *Adv. Phys.* **5**, 1 (1956).

Slater, J. C., *Quantum Theory of Molecules and Solids*, Vol. II, McGraw-Hill, New York, 1965.

INDEX